Fundamentals of
Farm Management

Fundamentals of Farm Management

Heidi Pratt

Larsen & Keller
www.larsen-keller.com

Fundamentals of Farm Management
Heidi Pratt
ISBN: 978-1-64172-606-1 (Hardback)

Larsen & Keller

Published by Larsen and Keller Education,
5 Penn Plaza,
19th Floor,
New York, NY 10001, USA

Cataloging-in-Publication Data

Fundamentals of farm management / Heidi Pratt.
    p. cm.
Includes bibliographical references and index.
ISBN 978-1-64172-606-1
1. Farm management. 2. Agricultural landscape management. 3. Agriculture.
4. Agricultural systems. I. Pratt, Heidi.
S561 .F86 2022
630.68--dc23

For more information regarding Larsen and Keller Education and its products, please visit the publisher's website www.larsen-keller.com

TABLE OF CONTENTS

Permissions

Index

It is with great pleasure that I present this book. It has been carefully written after numerous discussions with my peers and other practitioners of the field. I would like to take this opportunity to thank my family and friends who have been extremely supporting at every step in my life.

A farm is an area of land that is primarily devoted to agricultural processes with the primary goal of producing food and other crops. Farm management refers to the making and implementation of decisions that are involved in operating and organizing a farm to ensure maximum profit and production. The management strategies and techniques generally depend upon the type of farming and the size of the farm. The management of farms includes the usage of principles and concepts from the fields of plant and animal sciences in order to make informed decisions regarding soils, seeds and fertilizers. It also uses agricultural engineering for information on farm buildings, machinery and erosion control systems. A few of the different branches of farm management are poultry farm management, dairy farm management and livestock management. It aims to shed light on some of the unexplored aspects of farm management. Some of the diverse topics covered herein address the varied branches that fall under this category. This book will provide comprehensive knowledge to the readers.

The chapters below are organized to facilitate a comprehensive understanding of the subject:

Chapter – Introduction

Making and implementing decisions related to the running of a farm in order to maximize production and profit is termed as farm management. Some of the principles which are used in farm management are the law of marginal returns, law of substitution and law of combining enterprise. All these principles of farm management have been briefly introduced in this chapter.

Chapter – Livestock Farming and Management

The raising of domesticated animals for producing commodities such as eggs, meat, milk, etc. as well as labor is known as livestock farming. Livestock management deals with the activities related to the running of poultry farms, dairy farms and cattle ranches. This chapter discusses in detail these aspects and activities related to the farming and management of livestock.

Chapter – Poultry Farming and Management

The form of animal husbandry which deals with raising domesticated birds for eggs or meat is known as poultry farming. Poultry management deals with the production practices for maximizing the efficiency of production as well as handling poultry diseases. The topics elaborated in this chapter will help in gaining a better perspective about the different practices associated with poultry farming and management.

Chapter – Dairy Farm Management

The form of agriculture which focuses on the long-term production and/or processing of milk is known as dairy farming. Some of the technologies which are associated with dairy farming are rotolactor, bulk milk cooling tank and mixer-wagon. The diverse applications of these technologies in the management of a dairy farm have been thoroughly discussed in this chapter.

Chapter – Farm Equipments and Tools

Some of the tools and equipments which are used in farming are tractors, seed drills, harrows, rollers, cultivators, mowers, balers, hay rakes, tedders and combine harvesters. This chapter has been carefully written to provide an easy understanding of the varied applications of these equipments and tools in farming.

Heidi Pratt

Introduction

Making and implementing decisions related to the running of a farm in order to maximize production and profit is termed as farm management. Some of the principles which are used in farm management are the law of marginal returns, law of substitution and law of combining enterprise. All these principles of farm management have been briefly introduced in this chapter.

FARM MANAGEMENT

Farm management is the making and implementing of the decisions involved in organizing and operating a farm for maximum production and profit. Farm management draws on agricultural economics for information on prices, markets, agricultural policy, and economic institutions such as leasing and credit. It also draws on plant and animal sciences for information on soils, seed, and fertilizer, on control of weeds, insects, and disease, and on rations and breeding; on agricultural engineering for information on farm buildings, machinery, irrigation, crop drying, drainage, and erosion control systems; and on psychology and sociology for information on human behavior. In making his decisions, a farm manager thus integrates information from the biological, physical, and social sciences.

Because farms differ widely, the significant concern in farm management is the specific individual farm; the plan most satisfactory for one farm may be most unsatisfactory for another. Farm management problems range from those of the small, near-subsistence and family-operated farms to those of large-scale commercial farms where trained managers use the latest technological advances, and from farms administered by single proprietors to farms managed by the state.

In Southeast Asia the manager of the typical small farm with ample labour, limited capital, and only four to eight acres (1.6–3.2 hectares) of land, often fragmented and dispersed, faces an acute capital–land management problem. Use of early maturing crop varieties; efficient scheduling of the sequence of land preparation, planting, and harvesting; use of seedbeds and transplanting operations for intensive land use through multiple cropping; efficient use of irrigation and commercial fertilizer; and selection of chemicals to control insects, diseases, and weeds—all of these are possible measures for increasing production and income from each unit of land.

In western Europe the typical family farmer has less land than is economical with modern machinery, equipment, and levels of education and training, and so must select from the products of an emerging stream of technology the elements that promise improved crop and livestock yields at low cost; adjust his choice of products as relative prices and costs change; and acquire more land as farm labour is attracted by nonfarm employment opportunities and farm numbers decline.

On a typical 400-acre (160-hectare) corn-belt farm in the United States with a labour force equivalent to two full-time men, physical conditions and available technologies allow a wide range of options in farming systems. To reach a satisfactory income requires operating on an increasing scale of output and increasing specialization. Corn and soybean cash-crop farming systems have increased in number along with corn-hog-fattening farms and corn-beef-fattening farms. Thus, the choice of a farming system, the degree of specialization to be chosen, the size of operation, and the method of financing are top concerns of management.

For a typical crop-livestock farm in São Paulo's Paraíba Valley, Brazil, large-scale use of hired labour creates a substantial management problem. With 30 to 40 workers per establishment, procuring and managing the labour—keeping abreast of demand and supply conditions for hired labour, working out contractual arrangements (wage rates and other incentives), deciding how to combine labour with other inputs, and supervising the work force—are of critical importance.

A rancher with thousands of acres, whether in the pampas of Argentina, the plains of Australia, or the prairies of the United States, is concerned about the rate of increase of the herd through births and purchases and herd composition—cows, calves, yearlings, steers, heifers. Risks from drought, winter storms, and price changes can be high. Weather, prospective yields, and the price outlook are the constant concern of competent and alert farm managers.

On a collective farm in the Soviet Union with 30,000 acres (12,000 hectares) and 400 workers, major management decisions are made by party–state representatives; the collective-farm chairman responds largely to their directives, though the farm manager is being given greater autonomy. Major management concerns are determining optimal size of the collective, improving labour incentives, increasing crop and livestock yields, and reducing unit costs—with emphasis on levels of fertilizer, on pesticide and herbicide use, and on conservation of soil and water in crop production.

Thus, the character of the world's agriculture is shaped as millions of farmers manage the resources under their control in ways to obtain as much satisfaction as possible from their decisions and actions, which are made in a large variety of settings in regard to human, capital, and land resource combinations; technological possibilities; and social and political arrangements. Future agricultural progress depends on improving the quality of management and the environment in which farmers make decisions and on helping them adjust their decisions to the changing environment. In the low-income agricultures of the world in the 1980s, expanded research, improved input supplies and transport facilities, enlarged market opportunities, and an otherwise encouraging environment promise to open up a much wider area for managerial choice and decision making.

Basic Concerns

Land, Livestock, and Labour

A good farm manager is familiar with the legal description of the farm property for which he is responsible, location relative to other property, roads, markets, and sources of supply, the details of the field arrangement and farmstead layout, the farm's capital position or relation of debts to assets, and the resources of the farm, such as the capabilities of its soils. Such facts enable the manager to analyze and evaluate his resources and plan their use. To calculate profit potential, the farm manager estimates the yield expected from each acre or hectare of land and from each head of livestock. He then applies money prices to these quantities.

The size of a farm business, an indication of its profit-making potential, is measured by the total number of acres or hectares in the farm, acres or hectares planted to cash crops, productive man–work units (the number of workdays of labour required under average efficiency to care for crops and livestock), livestock units kept, capital invested, and total cash receipts. While total acreage is often used to describe farm size, it is not a very satisfactory measure since it does not specify how much land is hilly, stony, swampy, or otherwise unproductive. Total cropped land, total receipts, invested capital, or productive work units are better measures. Though livestock are counted by the head for the sake of comparison, for management purposes one cow is roughly equal in value to two calves, five hogs, 10 young pigs, seven sheep, 14 lambs, or 100 laying hens.

While the amount of land in a farm is more or less fixed, many farmers buy or rent additional acreage to increase their volume of output as a means of reducing unit costs. If such acreage is available within a reasonable distance, then land can often be profitably exploited. Other ways of increasing volume include bringing unimproved pasture and woodland into the cropping plan and shifting either to more intensive methods of cultivation or to more valuable crops. Before making major changes, the farm manager attempts to assure himself that the new crops will grow well and will find a market in his area. Almost all the governments of the world today have departments or ministries of agriculture which have been established for the purpose of advancing agricultural welfare by spreading technological information. Often these agencies perform extensive experimentation with new crop varieties, new cultivation techniques, and improved breeds of livestock, thus reducing the burden of risk upon the individual farm manager contemplating such changes. Considerable experimentation and research are also carried out by private agricultural supply firms that hope to improve their competitive position in the marketplace by developing a valuable new product.

In some of the developing countries, traditional patterns of land tenure and laws of inheritance may result in one farmer holding many quite small plots at some distance from each other. To reduce the resulting labour inefficiency and low productivity and to spur development of large-scale agriculture, governments in these countries have frequently legislated to permit or compel consolidation of such holdings.

Some kinds of farm work are directly productive, some are indirectly productive, and some are not productive at all. Work such as plowing, planting, cultivating, harvesting, feeding, and milking is directly productive. Maintenance of fences, buildings, and machinery, though often necessary, is not directly productive. Such work as trimming shrubbery and mowing lawns, unless it adds to the market value of the farm, is not considered productive. Similarly, capital can be highly productive, as in the case of livestock; indirectly productive (e.g., tractors, buildings, and supplies); or unproductive, as a large, showy barn or house. Land, too, can be highly productive, moderately so, or waste. Analysis of farm records has shown that farmers often overequip their property, thus using buildings and machinery to less than full capacity. Generally speaking, small farmers have been shown to have a higher proportion of their total investment in buildings than in machinery. In the developing countries, where relatively large quantities of human labour and relatively small amounts of capital are employed, a rather different problem exists. In these areas, farm managers need large numbers of people to work the fields during planting and harvest and far smaller numbers to perform routine cultivation tasks. In consequence, these countries face a problem of underemployment of agricultural labour during much of the year.

Financial Management and Large-scale Operation

The financial tools a farmer can use to analyze, plan, and control his business include financial statements, profit and loss statements, and cash-flow statements. A financial statement tells the amount of money invested in farm assets, outstanding debts, the owner's equity in the business, and the degree to which the farm is liquid and solvent. Liquidity is the ability to meet financial obligations on time, whereas solvency is the ability to pay all debts if the business is forced to discontinue. A profit and loss statement shows sources and amounts of income and operating expenses. Comparison of profit and loss statements over a period of years tells which resources have been most profitable and whether there has been an advance or decline in net income. A cash-flow statement shows the sources and uses of funds at given periods during the year. Such a statement provides a useful check on the accuracy of the farm's other business records.

For the traditional farmer, land and labour (his own and that of his family) are the major resources. Under favourable conditions, the farmer has changed his role from labourer to operator-manager; much larger farm units with high capital investments have resulted. Such conditions include the existence of a considerable body of applicable scientific knowledge, an opportunity for greater efficiency from large-scale operations, the existence of good markets and transportation, the opportunity to routinize and centrally direct farm work, and an absence of community antagonism to large-scale agriculture.

The trend to the substitution of capital for labour is especially noticeable in the United States, for example, where capital accounts for a steadily increasing proportion of farm inputs. In the United States in 1940, capital comprised 29 percent of farm inputs, labour 54 percent, and land 17 percent; by 1976 capital accounted for 62 percent of farm inputs, labour 16 percent, and land 22 percent. Capital typically replaces labour when large machines do the work of several men using smaller implements; when chemicals replace the scythe and hoe for weed control; when milking parlours, pipelines, and bulk tanks replace handmilking operations; when a mechanized installation replaces the fork and bushel basket in dairy, beef, or hog feeding; when automated sprinklers bring irrigation water to crops; when cisterns and lagoons handle animal waste; when combines and forced-air crop drying speed the harvesting of small grain; and in similar substitutions.

The technical knowledge that a modern large-scale farm manager must possess is frequently held to be far greater than that required of most businessmen with equal investment; the capital required to operate such a farm is beyond the reach of many. In consequence, financial-management techniques resembling those of industry are often employed. Capital is imported from the outside; production is scheduled to meet quantity, grade, and timing requirements; and labour is given specific tasks, as in a factory.

Recognizing the economic benefits of large-scale agriculture, many underdeveloped countries have attempted to create conditions for its existence. National governments, often with outside help, have financed large-scale development programs, involving irrigation or improvement of huge acreages by means of dams, drainage facilities, and canals, and these have revolutionized the lives of many traditional farm managers within the space of a few years. Improvements in crops and livestock, marketing techniques and organization, and transport and power have in some cases increased agricultural productivity and income several times over. Since capital and management have been in the hands of government, the traditional farm manager has, however, often lost some of his independence, and not all such programs have succeeded. Poor planning and management by government authorities and resistance from the farmers themselves have led to some expensive failures.

Reducing Market Risks

The marketplace for agricultural commodities is exceptionally risky for three important reasons. First, no single farm producer can place or withhold enough of a single item on the market to affect the market price; second, the quantity of a commodity taken off the market does not increase in proportion to price declines; third, the farm manager cannot respond to falling prices by quickly switching production from an unprofitable item to a profitable one. To reduce his risks and safeguard profits, the farm manager may specialize or diversify depending on conditions; he may also use the futures market.

A specialized farm manager concentrates his effort on the production of one item such as wheat, cotton, milk, eggs, or fruit. By such specialization he can realize the benefits of large-scale production and can make the most money from an enterprise in which he is highly skilled. On the other hand, the specialist is vulnerable to sudden changes in the market, to plant and animal diseases, and to soil exhaustion resulting from cultivation of a single crop.

Diversification—the spreading of one's talents over more than one farming enterprise—may be accomplished horizontally or vertically. Horizontal diversification means the production of more than one item for sale. In vertical diversification, the farm manager handles raw products after harvest by processing, packaging, transporting, or even selling at retail. A poultry farmer who produces eggs and washes, candles, grades, packages, and markets them at retail is said to be vertically diversified. He has taken on some of the jobs that could have been performed elsewhere, and as a result he generally receives a better return for his efforts.

Programs of agricultural diversification have been carried out by some developing countries, with the government acting as a kind of national farm manager. Upon achieving independence, nations such as Ghana and Nigeria, in West Africa, found their economies highly dependent upon a single raw agricultural export (cocoa for Ghana; palm oil for Nigeria). Sharply falling prices for these commodities or epidemics of plant disease were seen to have disastrous effect on national prosperity. Erosion problems also caused concern. The governments responded by horizontally diversifying into other profitable crops and vertically diversifying in the establishment of industries to process these commodities or turn them into manufactured goods before export.

A capable farm manager may use the futures market to try to minimize his risks. In the futures market, the farm manager contracts with a buyer to deliver a given quantity of some commodity at a specified date in the future for an agreed price. The buyer is often a speculator who hopes that prices will rise, enabling him to sell the commodity or the contract at a profit. Futures markets enable the farm manager to establish in advance a price for a crop or earn payment for holding a crop in storage. Futures markets also permit some farmers to speculate on a price increase without storing a crop, establish in advance the price of livestock feed intended for later use, and establish an advance price for livestock.

Special Concerns of Scale

Farm management specifics vary all over the world; it is possible here to cite only some of the most typical practices in several leading agricultural countries.

Large-farm Management

Research has shown that large farms produce more efficiently than small farms. In sugarcane production, for example, the most efficient farm may include many thousands of acres or hectares. Yet, a well-managed dairy farm might achieve greatest efficiency with two men and fewer than 100 cows. In the future, as technology advances, the farms that are managed most efficiently will probably be larger than the most efficient farms at present.

Large farms can reduce costs by claiming volume discounts on their purchases. They can negotiate prices on fertilizer, seed, crop chemicals, petroleum products, machinery, and repair services. Large operators also have an advantage in selling their products. Managers of large corn farms, for example, can contract directly with a large processor for an entire year's production of given quantity and quality for a specific date in the future, thus commanding a higher price. The middleman is eliminated, and production, handling, and processing can be prescheduled for greater efficiency. Large farms also have a smaller investment in machinery and buildings per crop acre.

PRINCIPLES USED IN FARM MANAGEMENT

Law of Diminishing Marginal Returns

This law states that: "An increase in the capital and labour applied to the cultivation of land causes in general a loss than the proportionate increase in the amount of produce raised unless it happens to coincide with an improvement in the art of agriculture."

There are three stages of the law of diminishing returns. They are: 1) stage II and 3) stage III. The positions of the parameters i.e. TP (Total Product), AP (Average Product), MP (Marginal Product) and EP (Elasticity stages of production are as under.

Stage I: Irrational Zone

1. This stage starts from origin and ends where AF & MF curves intersect each other.

2. The TP is increasing at increasing rate at first then at decreasing rate.

3. PP and MP both increase but MP is greater than IP.

4. The EP is greater than 1 (one).

Stage II: Rational Zone

1. It starts where PP & MP intersect each other and EP = 1. It ends when MP = 0.

2. TP increases but at decreasing rate.

3. MP starts to decline continuously and AP also starts to decline but it is greater than MP.

4. The elasticity of production (EP) is greater than zero but less than 1.

Stage III

1. This stage starts when MP is zero and TP is at maximum.

2. TP starts to decline and it declines continuously.

3. MP becomes negative, remains positive.

4. EP is always less than zero.

Law of Equimarginal Returns

The law of Equimarginal returns is concerned with the allocation of the limited amount of resource among different enterprises. The law states that "profits are maximized by using a resource in such a way that the marginal returns from that resource are equal in all cases."

Law of Substitution or Principle of Least Cost Combination

The objective of profit maximization can be achieved by two ways, one by increasing output and other by minimizing the cost. The minimization of cost can be possible by deciding the use of more than one resource in substitution of other resources.

Objective of Factor-factor Relationship is Two Fold

1. Minimization of cost at a given level of Output.

2. Optimization of output to the fixed factors through alternative resource use combinations.

$$y = f(x^1, x^2, x^3, x^4 .... xn)$$

Y is the function of x^1 and x^2 while other inputs are kept at constant. The relationship can be better explained by the principle of least cost combination.

Principle of Least Cost Combination

A given level of output can be produced using many different combinations of two variable inputs. In choosing between the two completing resources, the saving m the resource replaced must be greater than the cost of resource added.

The principle of least cost combination states that if two factor inputs are considered for a given output the least cost combination will be such where their inverse price ratio is equal to their marginal rate of substitution.

Marginal Rate of Substitution

MRS is defined as the units of one input factor that can be substituted for a single unit of the other input factor. So MRS of x2 for one unit of x1 is:

$$MRS = \frac{\text{Number of unit replaced resource}(x2)}{\text{Number of unit added resource}(x1)}$$

Price Ratio

$$PR = \frac{\text{Cost per unit of added resource}}{\text{Cost per unit of replaced resource}}$$

$$\frac{\text{Price of } X_1}{\text{Price of } X_2}$$

Therefore the least cost combination of two inputs can be obtained by equating MRS with inverse price ratio.

i.e. $X_2 \times Px_2 = x_1 \times Px_1$,

This combination can be obtained by following algebraic method or Graphic method:

Isoquant Product Curve

Iso = equal and quant = quantity. An Isoquant represents the different combinations of two variable inputs used in the production of a given amount of output.

Properties of Isoquant:

1. They slope down ward to the right: If more of one is used less of another input will be employed at the given level of output.

2. They are convex to the origin.

3. Isoquant does not intersect: It is not possible to have different outputs from a single combination of inputs.

4. Slope of Isoquant represents the MRS.

Types:

1. Convex Isoquant (decreasing rate) Good substitution.

2. Straight line Isoquant → Perfect substitute.

3. Right angel: No substitution complement.

Iso-costline

An Iso-cost line indicates all possible combinations of two inputs which can be purchased with a given amount of investment fund (outlay).

Each combination of inputs has same total cost which includes the cost of two inputs. (X_1 and X_2) combined.

$$\text{Total cost} = PX_1 \times X_1 + PX_2 \times X_2,$$

Properties of Iso-cost Line

1. As total outlay increases, the Iso- cost line moves higher and higher away from the origin and vis- a-visa.

2. The Iso-cost lines are straight.

3. Slope of Iso-cost line represents price ratio i.e. PX_1 / PX_2 when x1 is taken on X axis and X_2 on y axis.

Iso-cline

It is a line passes through the points of equal slope or MRS on an Isoquant surface. With the input price ratio being constant for each Isoquant the MRS between the inputs is the same for each level of output.

Ridge Line

These are also called as border line. Ridge lines join the end points of Isoquants. The area within the ridge lines is rational region of production arid beyond that the two regions are irrational. Therefore these lines represent the limits of economic relevance.

Expansion Path

All the least cost combination points are joined to each other; the result is an expansion line. As such, MRS = PR.

Border Line

Line joining the end points of Isoquant.

Principle of Combining Enterprise

This principle is very important as it describes the product – product relationship. Here, instead of considering the allocation of inputs among enterprises, we discuses enterprise combination or product mix involving product relationship. Algebraically the relationship can be written as under:

There can be various relationships that can exist between enterprises or products:

1. Joint Product: Two or more than two products are produced in the same production process. Eg. Paddy and straw.

2. Complementary Productions: In this case relationship is directly proportionate. With the increase in one product there is also increase in other product. E.g. the cultivation of leguminous crop followed by cereals gives this relationship.

3. Supplementary Productions: In this case, increase in one product does not effect for each other or they are independent and if relationship is there it is supplementary. Eg. Crop production and dairy enterprise.

4. Competitive Relationships: Here two products are said to be competitive when increase one needed to be reduction in other product. Eg. Two cereal crops.

Determination of Optimum Production Combination by Graphic Method

Iso-revenue Curve

It is the line which indicates the different combinations of two products which gives the same amount of revenue or income.

Properties of Iso-revenue Line

It is always straight line because the output prices do not change with the quantity sold.

The position of Iso revenue line shows the magnitude of the total revenue. As total revenue increases, the line moves away from the origin and vis-a-visa.

The slope of Iso-revenue curve represents the price ratio of two competing products.

Law of Opportunity Cost

1. The opportunity cost is also called as alternative cost.

2. Opportunity Cost is the earning from the next best alternative sacrificed.

Law of Comparative Advantage

The concept of comparative advantage is associated with:

1. Resource productivity,

2. Cost of production of enterprise.

IMPORTANCE OF FARM MANAGEMENT

Proper management of farmland is vital for an investor to capitalize on the overall appreciation of the asset. Farming today is more than just producing crops, it requires farmers and landowners to address profitability, fertility, conservation, and tax issues to name just a few. The importance of a knowledgeable and professional farm manager is essential for maximizing the appreciation and income of investment farmland.

All farmland is not created equal and a customized farm management plan and oversight will align the interests of the farmer and landowner to optimize their return on investment (ROI). The key to proper farm management includes focusing on the following areas:

- Profitability
- Leasing

- Production
- Fertility
- Conservation
- Capital Improvements
- Additional Revenue Opportunities
- Insurance
- Taxes
- Communication

Overlooking just one of these key tasks can lead to a significant loss in the ROI or degradation of the farmland. Farming over the last decade has become one of the most profitable industries, although the improvement in economics has not necessarily flowed back to the landowners. Professional farm management services will not only allow investors to optimize their ROI, but own an asset that can be passed down for generations.

Profitability

Productive cropland is profitable on multiple levels by producing food to the growing population and also providing its owner with intermittent cash flow with stable appreciation upside. Farmland deserves to be a part of every well diversified investor's portfolio.

Farm management is essential for farmland owners to maximize annual ROI and long-term capital appreciation. Any farmland should increase in value and produce annual income to land owners, but with progressive farm management, landowners can expect much higher profitability.

Leasing

A key part of the farm manager's duties are their relations with the tenant operator. Choosing the appropriate operator can make or break an investment in farmland. The operator will help determine the short and long-term fertility, production, conservation, and cosmetics of a property. Even one mismanaged year of farming can cause significant damage to a property.

To ensure the right tenant is chosen, managers will interview many qualified farmers, including a thorough inspection of the tenant's operation. Background checks are also important as lenders and local contacts will give a better insight to the credit worthiness of the farmer.

The manager should have a large pool of potential tenants that are competing for a given property. Competition between operators for leasing a property will give the landlord the highest possible rent. A good manager will know what market rent is in the area and use that as a starting point for negotiating.

Identifying the right type of lease will differentiate a good farm manager from a great farm manager. Each lease presents different cash flow and risks. It is imperative that the owner and manager are on the same page and comfortable with the chosen lease and the amount of risk exposure to the landlord.

There are four commonly used leases:

- Cash Rent Lease: Tenant pays a specified amount of cash per acre per year to farm the property. The payment is made in full before any seed is planted; typically paid on March 1st of the leased year.

- Flex Lease: This lease is a variation of the Fixed Cash Lease. Rent will vary based on yields and crop prices throughout the year. This lease gives the landlord a share of the risks associated with farming in a given year.

- Crop-share Lease: Tenant pays landlord with a percent of the crop or income that is produced. In a typical lease, the landlord would split the input farming costs of fertilizer, seed, pesticides, etc. In some instances, the landlord is responsible for marketing their share of the crop as well.

- Custom Farm Lease: This is the most involved an owner can be in a farming operation without actually farming. The owner takes on the input costs including fertilizer, seed, pesticides, herbicides, etc. while an operator is contracted out by the owner for performing farming tasks throughout the year. 100% of the production is retained by the owner.

Although each lease has its differences, there are two important aspects in common; price and length.

Price of rent can be the deciding factor on whether or not to purchase a property for investment. A good manager will not only use market averages, but also forecast an income statement for the property to estimate what the owner's share of income should be.

The second part of negotiating a lease is the length. A long-term lease is not always in the landlord's best interest. If a lease is signed for $100 per acre for five years, and after two years the market rent increased from $100 to $125, the landlord is missing out on $25 per acre. Negotiating a one to three year lease ensures the landlord will be receiving the current fair market rent.

Production

It is always important to track crop conditions, but for certain leases, including flex and custom farming where the landlord has upside potential in the production of the property, crop conditions are of utmost importance. Farm managers work with farmers to ensure planting was successful and the correct seed varieties were planted for the climate forecasted during the upcoming growing period.

Throughout the planting season, farm managers keep in contact with operators to note how the crops are progressing which will help build a strong historical file for the property. Farmland with consistent proven yields of 200 bushels of corn will have a higher value than a property that has a volatile production history of similar soil quality as future buyers prefer consistent yielding properties.

Harvest will produce yield data that farm managers record for the property's historical file. Often operators will have a yield map that will be supplied to the farm manager, helping the manager understand where the strongest yielding areas of the field are located along with other features like compaction or poor drainage. Matching soil types from a soil map to the yield map often will reveal where the poorly drained areas of the property are located. A manager will then explore where additional drainage relief is needed, let it be tile, surface intakes, or a waterway.

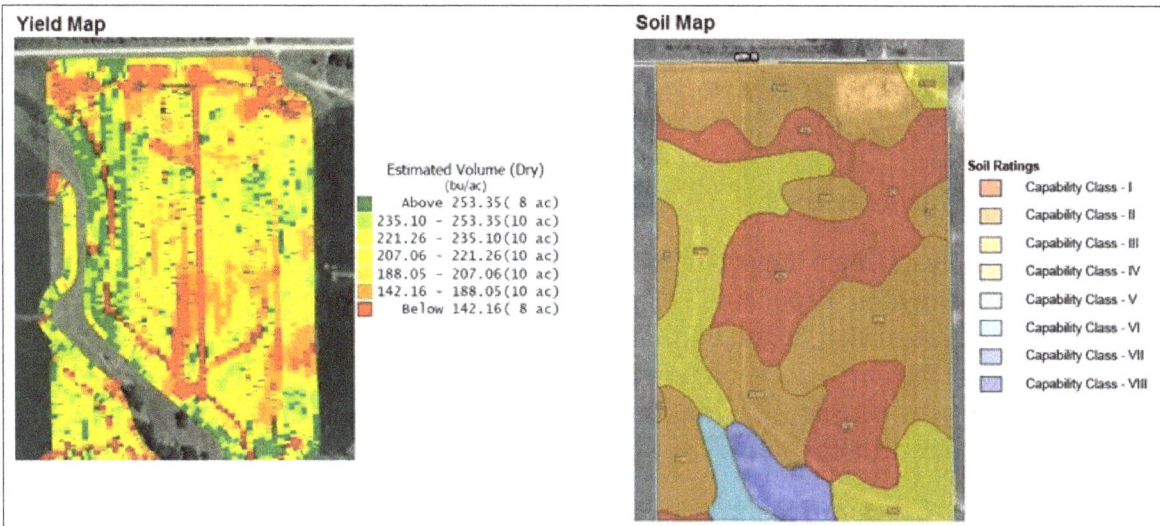

Yield Map / Soil Map

Fertility

Once harvest is complete and farmers begin to plan their input purchases for the following year, farm managers will work with farmers to gauge the fertility of the property with the use of soil samples. Farmers test soil fertility via soil samples at least every other year to make sure the correct amount of fertilizer is used to achieve optimum yields.

Additionally, farmers do not want to apply too much fertilizer than their soil Cation-Exchange Capability (CEC) can handle which would lead to wasted fertilizer and money. Soil CEC refers to the amount of nutrients a soil can absorb efficiently at a given pH level. If a soil has a low CEC, then over fertilizing can lead to fertilizer runoff and waste.

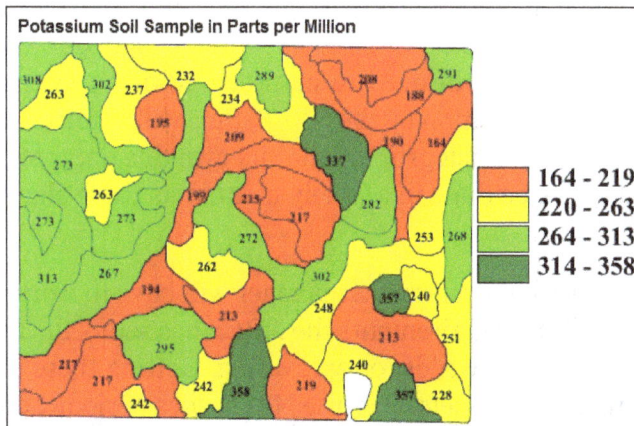

Potassium Soil Sample in Parts per Million

Farm managers work with the farm operator after comparing yields maps, soil maps, and soil sample maps to discuss the property's nutrient program moving forward to meet fertility goals while maximizing yields.

Tracking soil fertility through soil samples is essential for future property value increases. Typically, purchasers do not want to buy a property with poor nutrient levels. Although poor fertility is not typically permanent, application of macro and micro nutrients along with lime are required to bring fertility up to adequate levels over a period of multiple years which can carry significantly high costs and lead to lower rents. If a landowner currently holds a property with poor fertility, it is important to have a farm manager work with the operator to build a nutrient program to rebuild the property's fertility.

Conservation Management

Farmland is arguably the most important asset to sustaining life on an ever growing planet. World population is growing at an exponential rate, and taking care of and conserving farmland is essential to feeding the growing population. In order to conserve our precious asset, the USDA developed the Natural Resource Conservation Service (NRCS). The NRCS helps land owners reduce soil erosion, enhance water supplies, improve water quality, increase wildlife habitat, and reduce damages from floods and other natural disasters.

There are a variety of programs a farmland owner could sign up for; the programs most used by farmland owners include the Conservation Reserve Program (CRP), Grassland Reserve Program (GRP), Water Bank Program (WBP), Wetland Reserve Program (WRP), and Emergency Watershed Program (EWP). Each has its own unique characteristics, but in general when land is enrolled in these programs, the owner receives yearly payments and the contracts typically last 10 to 15 years.

In order to sign up for these programs, the farm manager must put in a considerable amount of work and due diligence. Having a good relationship with the local NRCS office can help simplify signing up for a conservation program. Depending on when a landowner starts to sign up for a program, communication with the NRCS office will last for one to two months. Weekly interactions via phone or e-mail are a must and having a good relationship with an NRCS representative is essential for a smooth process.

To start the process of signing up for a conservation program, a manager must first have detailed knowledge of the land and what areas they would like to have signed into a program. Aerial maps showing outlines of possible areas will need to be procured and presented to the NRCS. Upon review of maps and physical examination by the NRCS, they will determine what type of conservation program the land falls under. Each program has its own due dates for signing up. It is crucial that the work and due diligence be done in a timely manner as these sign ups typically only come once a year.

Upon approval of the conservation program, a manager will have to work with the current operator to make sure they meet all the requirements the NRCS has set forth. This may include seeding grass, plugging drains, spraying for weeds, or excavating. Getting these requirements completed correctly and in a timely manner is essential as the NRCS does periodic, onsite, evaluations. If the requirements are not met or part of the agreement is breached, it could result in fines or payment stoppage. This should not happen if the manager has done their due diligence and seen the process through to the end.

Adding Value through Capital Improvements

In order to maximize appreciation, ideal farmability should be targeted by the farm manager and owner which will include capital improvement projects. By taking a diverse approach to capital improvements, farm managers can present projects that can fit any landowner's budget. Common capital improvements cover drainage, erosion, and access.

Adding drain tile is one of the single best additions one can make to a property. The primary reason to install drainage tile in a farm field is to increase productivity through healthier crops. Ideal soil is made up of 50% soil, 25% water, and 25% air. When a heavy rain elevates the water table, the soil loses its 25% air make-up, which will hurt crop growth and increase soil pH levels. Fixing drainage issues by installing drainage tile typically pays for itself through increased yields within five years of installation by farmers paying higher rent. Tile projects have a wide range of cost from small localized projects totaling $1,000 to $2,000 to large scale parallel pattern tile projects running upwards of $750 per acre.

The cost of adding drainage tile can often be immediately added on to a property's value. Since drainage tile is eligible for accelerated depreciation, farm managers work with accountants, farmers, excavators, and previous owners to assign a value to any tile in a property so the landowner can write off the cost as a 100% tax deduction. An accurate tile depreciation valuation can save landowners tens of thousands of dollars on taxable income. Additionally, working installation of drainage tile into a lease can provide opportunity for a landowner to get discounted tile if their operator installs the tile at a reduced rental price.

Precious soil can erode away via the wind, rain, or other weather elements thus striping a landowner of their asset. Farm managers are aware of erosion issues and should be constantly monitoring every property for signs of erosion so they may act fast to limit any soil loss. Simple solutions to localized erosion including installing a berm or retaining wall and extreme measures including installation of waterways, ditches, terraces, or other major excavation work which would be administered by the farm manager.

Erosion Before Management   Erosion After Management

Another excellent way to increase property value is by adding field access points. On a yield map managers will note poor yields that were caused by soil compaction. Often soil compaction is caused by heavy machinery running over the same area repetitively; often near field access points. By adding multiple points of access, farmers can cut down their traffic on compaction areas, thus increase total production over time.

In years of drought, any farmland would benefit from an irrigation system. Farm managers will work with irrigation outfitters to price an irrigation system and generate a long-term economic analysis of installing an irrigation system. Depending on the property's location and soil makeup, irrigation can substantially increase yields, cash rents, and position the property for ultimate appreciation.

Additional Revenue Opportunities

Revenue can be created on farmland, not just via farming operations, but also through other means including wind easements, hunting rights, and advertisements.

Landowners can benefit from having a wind farm a part of their property by leasing the property's wind rights. These contracts are created in the first process of building a wind farm, so land owners get paid prior to any building. When the wind farm is finalized and constructed, land owners will receive fixed and variable payments based on electricity production. Landowners could receive up to $15,000 per year on a 160 acre parcel, although each wind company's contract will differ. Farm managers will handle agricultural impact and economic research behind any such wind project, keeping in mind the property's future for appreciation at all times.

Midwest farmland produces the best corn yields in the world, but also some of the best hunting as well. Upland birds, waterfowl, and deer hunting are some of the most sought after hunting experiences that outdoorsmen demand. By leasing out farmland to hunters for hunting rights, landowners can generate increased annual ROI on top of their crop production lease. Farm managers will source tenants, work with an attorney to draft a proper lease, recommend insurance protection, and manage the hunting rights lease on farmland.

Often farmland is located on a desolate gravel road upwards of 15 miles from the nearest housing development or town, but sometimes farmland will be located on a busy highway or interstate with high amounts of traffic. Roadways with high traffic levels lead way for advertisement potential on neighboring farmland. Progressive farm managers will investigate advertisement potential of property and work with a billboard company to put together an investment proposal to see if billboard advertisement would generate additional ROI via rent payments and increased land value.

Insurance

A good farm manager will go above and beyond standard management to insure a client has protection against liability. Farming can be a dangerous job with many hazards as large machinery will be used on a given property throughout the year. Accidents can happen, and making sure a client is protected from liability is important. Typically a farm manager will not administer insurance, but pointing the client in the right direction and helping with the process is a duty a manager should embrace.

Taxes

Property taxes are paid each year, due on specific dates that vary by state. A manager will need to be familiar with each state's due dates to ensure their client is compliant as clients might have multiple properties across several states. Failure to pay property taxes could result in fines on the property, so educating a client on property taxes is essential to a manager's duties.

Communication

Communicating with clients is vital throughout the management process to ensure the client and managers are in collaboration to meet shared goals. Communicating why certain duties are performed will keep all parties on the same page.

Typically, the farm manager will provide a written annual report, highlighting the previous farming season, the outlook for the coming season, and potential capital improvement projects. This will assist the client in documenting the farming operation and better understanding the economics of their investment.

ANIMAL HUSBANDRY

Animal husbandry or animal farming is branch of agriculture. Particularly, animal husbandry deals with management of animals i.e. animals are raised for acquiring essential products like milk, meat and eggs. Therefore, animal husbandry includes farming of animals like cow, goat, buffalo, fish, sheep, poultry, etc. More specifically, animal husbandry is divided into cattle farming, poultry farming, fish production and beekeeping.

Cattle Farming

In this type of animal husbandry cattles like cow, buffalo, bullocks, etc are farmed. Hence, Cow or Bos indicus, buffalo or Bos bubalis are farmed for milk, therefore, are also known as milch animals. Bullocks are used for transportation and labour hence they are called draught animals. Milk production in cattles depends on lactation period. Therefore, for increasing milk production there is need of increasing the lactation period. Particularly, foreign breeds like Jersey, Brown swiss are chosen for high milk production. Hence, for good production there is need of good health of cattles. Factors that provide good health is not only appropriate food but also care for cattles and prevention of cattles from various diseases.

Health of Cattles

Food and Care	Diseases of Cattle
Cattles must be provided with roughage having lots of fibres and concentrates with lot of vitamin.	External parasites causes skin diseases in cattles.
For high milk production, food must be rich in vitamins and minerals.	Internal parasites causes stomach and intestinal problems.
Protection of cattles from heat, rain and cold must be done by maintaining roof shades.	Moreover, bacteria and viruses causes infectious diseases.
Also, excess of water is shade must be avoided, hence, floors must be sloping.	Such diseases are prevented by vaccination.
Dirt and loose hair must be removed by brushing.	

Poultry Farming

Animal husbandry of poultry involves farming of birds for eggs and meat. Among the birds layers are raised for eggs and broilers are raised for meat. Therefore, for good production of poultry birds, birds must be provided with proper food. In particulars, layers must be provided with fibrous food whereas, broilers must be provided with vitamin, proteins and fat rich food. Also, feathering of poultry, temperature and cleanliness of farms must be maintained. Also poultry must be protected from diseases. This can be done by spraying disinfectants in farms and by proper vaccination of birds.

Fish Production

In animal husbandry, fish production involves breeding of fishes. Fishes are rich in protein and thus are used as food. Fishes are obtained from natural resources or also by fish farming.

- Marine shery: It has large coast line of 7500 km which provides area for fishing. Hence, Pomfret, Saradine, Bombay duck are caught using fishing net from the sea water. By mariculture, fishes like Mullets, Bhetki, Prawns, etc., are farmed in sea water.

- Inland shery: Fishing is done in river water, ponds, canals, brackish water, etc. It provides less yield and allows culture as well as capture fishing.

- Composite fish culture: In this case, five to six variety of different fishes are farmed in single pond. Catlas are surface feeder, Rohu feeds on middle zone, mrigal and common carps are bottom feeders. Moreover, grass carps feeds on weeds inside water. Therefore, all food provided to are evenly utilised by fishes. Problem with fish culture is that, fishes lay eggs in monsoon. Hence, to increase rate of reproduction so that fish can lay eggs frequently, fishes are provided with hormonal stimulation. By hormonal stimulation method fishes can reproduce any time.

Beekeeping

In this type of animal husbandry, bee is raised for production of honey and wax. Bees are raised in bee farms and apiaries. Some bees used for honey production are, Indian bee (Apis cerana indica), rock bee (Apis dorsata), little bee (Apis florae). Moreover, high yield of honey is obtained from Italian bee (Apis mellifera) because these bees collect more nectar from the flowers and sting less. Also, quality of honey depends on availability of flowers and pollens, whereas, its taste differs with availability of flower.

Livestock Farming and Management

The raising of domesticated animals for producing commodities such as eggs, meat, milk, etc. as well as labor is known as livestock farming. Livestock management deals with the activities related to the running of poultry farms, dairy farms and cattle ranches. This chapter discusses in detail these aspects and activities related to the farming and management of livestock.

Livestock farming is the raising of animals for use .An efficient and prosperous animal agriculture historically has been the mark of a strong, well-developed nation. Such an agriculture permits a nation to store large quantities of grains and other foodstuffs in concentrated form to be utilized to raise animals for human consumption during such emergencies as war or natural calamity. Furthermore, meat has long been known for its high nutritive value, producing stronger, healthier people.

Ruminant (cud-chewing) animals such as cattle, sheep, and goats convert large quantities of pasture forage, harvested roughage, or by-product feeds, as well as nonprotein nitrogen such as urea, into meat, milk, and wool. Ruminants are therefore extremely important; more than 60 percent of the world's farmland is in meadows and pasture. Poultry also convert feed efficiently into protein; chickens, especially, are unexcelled in meat and egg production. Milk is one of the most complete and oldest known animal foods. Cows were milked as early as 9000 BCE. Hippocrates, the Greek physician, recommended milk as a medicine in the 5th century BCE. Sanskrit writings from ancient India refer to milk as one of the most essential human foods.

Cattle

Beef Cattle Breeds

The British Isles led the world in the development of the principal beef breeds; Herefords, Angus, beef Shorthorns, and Galloways all originated in either England or Scotland. Other breeds of greatest prominence today originated in India (Brahman), France (Charolais; Limousin; Normandy), Switzerland (Simmental), and Africa (Africander). The Hereford breed, considered to be the first to be developed in England, probably descended from white-faced, red-bodied cattle of Holland crossed with the smaller black Celtics that were native to England and especially to Herefordshire. By the middle of the 18th century the slow process of selective breeding that resulted in the smooth, meaty, and prolific Herefords had begun. The United States statesman Henry Clay of Kentucky imported the first purebred Herefords to America in 1817.

The Hereford, which became the most popular beef breed of the United States, is distinguished by its white face, white flanks and underline, white stockings and tail, and white crest on the neck. Its body colour ranges from cherry to mahogany red. It is of medium size, with present-day breeders making successful efforts to increase both its rate of weight gain and mature size, in keeping with the demand for cheaper, leaner beef.

The Polled Hereford is a separate breed of cattle originating from hornless mutations in 1901. It has the same general characteristics as the horned Hereford and has gained substantial favour because of its hornlessness and often faster rate of weight gain.

The Aberdeen Angus breed originated in Scotland from naturally hornless aboriginal cattle native to the counties of Aberdeen and Angus. Solid black, occasionally with a spot of white underneath the rear flanks, the breed is noted for its smoothness, freedom from waste, and high quality of meat.

Although the native home of the Galloway breed is the ancient region of Galloway in southwestern Scotland, it probably had a common origin with the Angus. The Galloway is distinguished by its coat of curly black hair. Though the breed has never attained the prominence of other beef breeds, it has been used extensively in producing blue-gray crossbred cattle, obtained by breeding white Shorthorn bulls to Galloway cows.

The beef, or Scotch, Shorthorn breed developed from early cattle of England and northern Europe, selected for heavy milk production and generally known as Durham cattle. These were later selected for the compact, beefy type by the Scottish breeders. Emphasis on leaner, high quality carcasses in the second half of the 20th century has diminished the popularity of this breed. The Polled Shorthorn originated in 1888 from purebred, hornless mutations of the Shorthorn breed. The milking, or dual-purpose, Shorthorn, representing another segment of the parent Shorthorn breed, also was developed in England to produce an excellent flow of milk as well as an acceptable carcass, therefore resembling the original English type of Shorthorn. Shorthorns range in colour from red through roan, to white- or red-and-white-spotted.

Shorthorn bull

The Brahman breed originated in India, where 30 or more separate varieties exist. Preference is given to the Guzerat, Nellore, Gir, and Krishna Valley strains, which are characterized by a pronounced hump over the shoulders and neck; excessive skin on the dewlap and underline; large, droopy ears; and horns that tend to curve upward and rearward. Their colour ranges from near white through brown and brownish red to near black. Their popularity in other areas such as South America and Europe, into which they have been imported, is attributable mainly to their heat tolerance, drought resistance, and resistance to fever ticks and other insects. The Santa Gertrudis was developed by the King Ranch of Texas by crossing Brahman and Shorthorn cattle to obtain large, hearty, tick-resistant, red cattle that have proved to be popular not only in Texas but in many regions along the semitropical Gulf Coast. Until the tick was eradicated in the southern and southwestern United States, Brahman crosses were raised almost exclusively there.

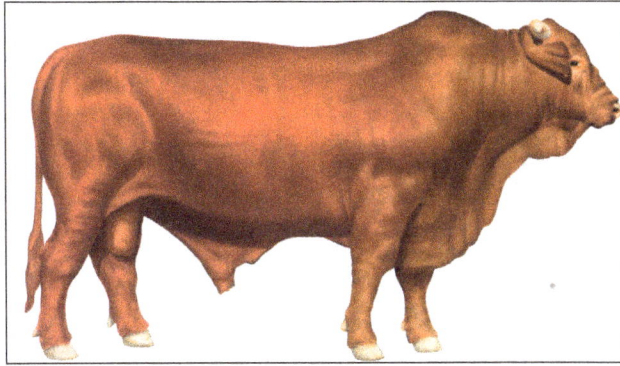

Santa Gertrudis bull

Several lesser breeds have been developed from crosses of the Brahman on other beef breeds such as: the Charbray (Charolais), Braford (Hereford), Brangus (Angus), Brahorn (Shorthorn), and Beefmaster (Brahman-Shorthorn-Hereford).

The Charolais breed, which originated in the Charolais region of France, has become quite popular in the United States for crossing on the British breeds for production of market cattle. The superior size, rate of gain, and heavy muscling of the pure French Charolais and the hybrid vigour accruing from the crossing of nonrelated breeds promise an increased popularity of this breed. Many American Charolais, however, carry significant amounts of Brahman blood, with a corresponding reduction in size, rate of gain, and muscling. Important in France, the Charolais is the foremost meat-cattle breed in Europe.

The Limousin breed, which originated in west central France, is second in importance to the Charolais as a European meat breed. Limousin cattle, often longer, finer boned, and slightly smaller than the Charolais, are also heavily muscled and relatively free from excessive deposits of fat.

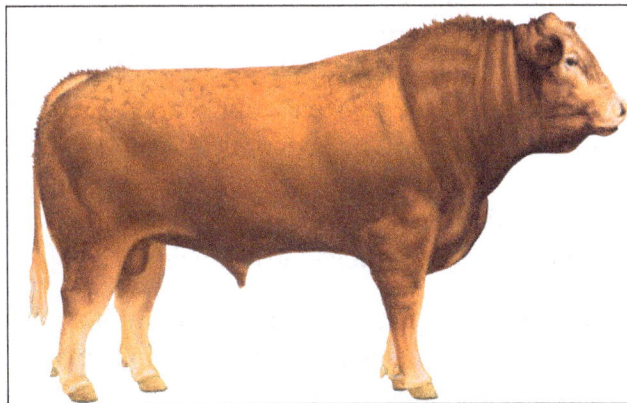

Limousin bull

The most prevalent breed of France, the Normandy, is smaller than the Charolais or Limousin and has been developed as a dual-purpose breed useful for both milk and meat production. A fourth important breed is the Maine–Anjou, which is the largest of the French breeds.

The Simmental accounts for nearly half of the cattle of Switzerland, Austria, and the western areas of Germany. Smaller than the Charolais and Limousin, the Simmental was developed for milk, meat, and draft. It is yellowish brown or red with characteristic white markings.

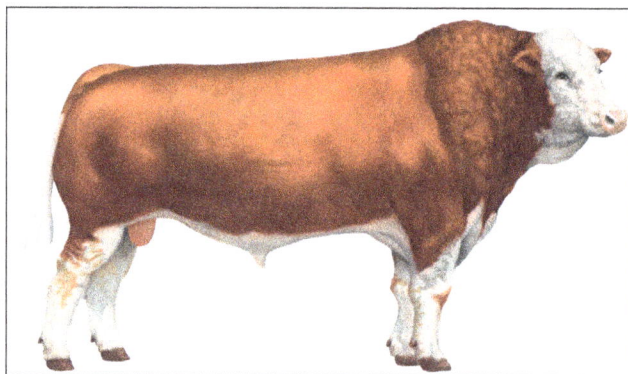

Simmental bull

Beef Cattle Feed

Beef cattle can utilize roughages of both low and high quality, including pasture forage, hay, silage, corn (maize) fodder, straw, and grain by-products. Cattle also utilize nonprotein nitrogen in the form of urea and biuret feed supplements, which can supply from one-third to one-half of all the protein needs of beef animals. Nonprotein nitrogen is relatively cheap and abundant and is usually fed in a grain ration or in liquid supplements with molasses and phosphoric acid or is mixed with silage at ensiling time; it also may be used in supplement blocks for range cattle or as part of range pellets. Other additions to diet include corn (maize), sorghum, milo, wheat, barley, or oats. Fattening cattle are usually fed from 2.2 to 3.0 percent of their live weight per day, depending on the amount of concentrates in the ration and the rate at which they are being fattened. Such cattle gain from 2.2 to 3.0 pounds (1.0 to 1.4 kilograms) per day and require from 1.3 to 3.0 pounds (0.6 to 1.4 kilograms) of crude protein, according to their weight and stage of fattening. Up until the early 1970s, when the practice was prohibited, fattening cattle were given the synthetic hormone diethylstilbestrol as a supplement in their feed or in ear implants. The use of this synthetic hormone results in a 10 to 20 percent increase in daily gain with less feed required per pound of gain. Synthetic vitamin A sources have become so cheap as to permit the use of 10,000 to 30,000 International Units per day for cattle being fattened for market (finished) in enclosures bare of vegetation (drylots) used for this purpose. The economics of modern cattle finishing encourages the use of all-concentrate rations or a minimum of roughage, or roughage substitutes including oyster shells, sand, and rough plastic pellets. Corn (maize) silage produces heavy yields per acre at a low cost and makes excellent roughage for beef-cattle finishing.

Beef cows kept for the production of feeder calves are usually maintained on pasture and roughages with required amounts of protein supplement and some grain being fed only to first-calf heifers or very heavy milking cows. Most beef cows tend to be overnourished and may become excessively fat and slow to conceive unless they happen to be exceptionally heavy milkers. Most pregnant cows go into the winter in satisfactory condition and need to gain only enough to offset the weight of the fetus and related membranes. They can therefore utilize coarser roughages, having a total daily crude protein requirement of from 1.3 to 1.7 pounds (0.58 to 0.76 kilograms). Daily vitamin A supplement at the rate of 18,000 to 22,000 International Units per cow is advisable unless the roughages are of a green, leafy kind and the fall pasture has been of excellent quality. Feed requirements for bulls vary with age, condition, and activity, from 2.0 to 2.4 pounds of crude protein per day; from 25,000 to 40,000 International Units of vitamin A; and during breeding periods nearly the same energy intake as calves or short yearlings being finished for market, the main feeding requirement being to prevent their becoming excessively fat.

All cattle require salt (sodium chloride) and a palatable source of both calcium and phosphorus, such as limestone and steamed bone meal. Most commercial salts carry trace minerals as relatively cheap insurance against deficiencies that occasionally exist in scattered locations.

Beef Cattle Management

Beef production has become highly scientific and efficient because of the high cost of labour, land, feed, and money. Most brood-cow herds, which require a minimum of housing and equipment, are managed so as to reduce costs through pasture improvement and are typically found in relatively large areas and herds. Other aspects of management include performance testing for regular production of offspring that will gain rapidly and produce acceptable carcasses and the use of preventive medicine, feed additives, pregnancy checks, fertility testing of sires, artificial insemination of some purebred and commercial herds, protection against insects and parasites, both internal and external, adequate but not excessive feed intakes, and a minimum of handling.

Calving of beef cows is arranged to occur in the spring months to take advantage of the large supplies of cheap and high-quality pasture forages. Fall calving is less common and occurs generally in regions where winters are moderate and supplies of pasture forage are available throughout the year. Calves are normally weaned at eight to ten months of age because beef cows produce very little milk past that stage and also because they need to be rested before dropping their next calf. Feeder calves sell by the pound, so that weight for age is even more important than conformation or shape. Consequently, crossbred cattle are used; their hybrid vigour results in greater breeding efficiency and milk production on the part of the dam, as well as greater birth weight, vigour, and gaining ability on the part of the offspring.

Beef cows are normally first bred at 15 to 18 months. The gestation period is 283 days, and the interval between estrus, or periods in which the dam is in heat, is 21 days. Cows should produce a living calf every 12 months. Pasture breeding, in which nature is allowed to take its course, calls for one mature bull for every 25 cows, whereas hand breeding, in which control is exercised by the breeder, requires half as many bulls. Artificial insemination permits one outstanding sire to produce thousands of calves annually.

Diseases of Beef and Dairy Cattle

Dairy cattle are susceptible to the same diseases as beef cattle. Many diseases and pests plague the cattle industries of the world, the more serious ones being prevalent in the humid and less developed countries. One of the more common diseases to be found in the developed countries is brucellosis, which has been controlled quite successfully through vaccination and testing. This disease produces undulant fever in humans through milk from infected cows. Leptospirosis, prevalent in warm-blooded animals and humans, is caused by a spirochete and results in fever, loss of weight, and abortion. Bovine tuberculosis has been largely eliminated; where it has not, it can infect other warm-blooded animals, including humans. Test and slaughter programs have proved effective. Rabies, caused by a specific virus that also can infect most warm-blooded animals, is usually transmitted through the bite of infected animals, either wild or domestic. Foot-and-mouth disease has been eliminated from most of North America, some Central American countries, Australia, and New Zealand. The rest of the world is still plagued by the disease, which attacks all cloven-footed animals. Humans are mildly susceptible to this organism. Successful vaccinations have been developed for blackleg, malignant edema, infectious bovine rhinotracheitis (or red nose), and several other diseases. Anaplasmosis, common to most tropical and semitropical regions, is spread by the bite of mosquitoes and flies. Anthrax, caused by a generally fatal bacterial infection, has been largely eliminated in the United States and western Europe. Rinderpest, still common to Asia and Europe, is caused by a specific virus that produces high fever and diarrhea. An infectious fever sometimes called nagana, caused by the tsetse fly, attacks both cattle and horses and is prevalent in central and southern Africa, as well as in the Philippines. Grass tetany and milk fever both result from metabolic disturbances. Bloat, caused by rapid gas formation in the rumen, or first compartment of the stomach, is sometimes fatal unless relieved. Pinkeye is an infectious inflammation of the eyes spread by flies or dust and is most serious in cattle having white pigmentation around one or both eyes. Mastitis, an inflammation of the udder, is caused by rough handling or by infection. Vibriosis, a venereal disease that causes abortion; pneumonia, an inflammation of the lungs; and shipping fever all cause serious losses and are difficult to control except through good management. Broad-spectrum antibiotics (antibiotics that are effective against various microorganisms), as well as powerful and specific pharmaceuticals, are effective and profitable means of keeping cattle herds healthy. Vermifuges, which destroy or expel parasitic worms, and insecticides, which kill harmful insects, are also highly effective and much used.

Pigs

Pigs are relatively easy to raise indoors or outdoors, and they can be slaughtered with a minimum of equipment because of their moderate size . Pigs are monogastric, so, unlike ruminants, they are unable to utilize large quantities of forage and must be given concentrate feed. Furthermore, pigs have only one primary economic use—as a source of meat (pork) and lard—unlike most other livestock, such as cattle and sheep, which have many other important economic uses.

Female pigs can have as many as 20 piglets in a litter.

Breeds

There are more than 300 known breeds or local varieties of pigs throughout the world. Following is a brief description of the better-known commercial breeds.

The Hampshire pig, which originated from the Norfolk thin-rind breed of England, is black with a white belt completely encircling its body, including both front legs and feet. There should be no white on the head or the ham.

Hampshire boar

The Yorkshire pig, which originated early in the 19th century in England, where it was considered a bacon type, is long, lean, and trim with white hair and skin. Found in most countries, this breed is probably the most widely distributed in the world.

Yorkshire (Large White) boar

The Duroc-Jersey breed originated in the eastern United States from red pigs brought by Christopher Columbus and Hernando de Soto. The modern Duroc, originated from crosses of the Jersey Red of New Jersey and the Duroc of New York in the late 19th century, ranges from golden-red to mahogany-red in colour, with no black allowed. This breed proved particularly suitable for feeding in the U.S. Corn Belt (parts of Ohio, Indiana, Illinois, Wisconsin, Minnesota, South Dakota, Nebraska, Missouri, and Oklahoma; all of Iowa) and has been extensively used in Argentina, Canada, Chile, and Uruguay. It is recognized for the quality of its meat.

Duroc boar

The Poland China originated about 1860 in southern Ohio from a number of different breeds common to that area. The Spotted Poland China originated in Indiana about 1915 from crosses of the Poland China and the native spotted pigs.

Poland China pig

The Chester White, which originated in Chester county, Pennsylbalia, after 1818, is restricted to the United States and Canada.

The Berkshire, which originated in Berkshire, England, about 1770. It is used for fresh pork production in England and Japan; a larger bacon type has been evolved in Australia and New Zealand. Like the Duroc breed, the Berkshire is noted for the quality of its meat.

The Landrace is a white, lop-eared pig found in most countries in central and eastern Europe, with local varieties in Denmark, Germany, the Netherlands, and Sweden. World attention was first drawn to the Landrace by Denmark, where since 1895 a superior pig has been produced, designed for Denmark's export trade in Wiltshire bacon to England and developed by progeny testing (the selection of boars for breeding on the basis of the scientific assessment of their progeny). Sweden also has progeny tested from Landrace stock but for a shorter period. Pigs from Sweden were first exported to England in 1953, when prices of up to £1,000 were paid. This resulted in a worldwide Landrace explosion, and most major pig-producing countries have since taken stock.

Landrace boar

The importance of the Asian pig breeds was recognized in the use of Chinese and "Siamese" pigs from southeastern Asia in the improvement of early European and North American breeds and is reflected in the name of the world-famous Poland China. China leads the world in pig numbers, and pork is traditional in the Chinese diet.

Daweizi pig

Breeding and Growth

Purebred production, or line breeding, is used to concentrate desired genes—for example, litter size or growth rate—within a population of animals. White pig breeds are generally noted for large litters (a maternal characteristic) and coloured breeds for rapid growth and meat quality (paternal characteristics).

Before 1980 most genetic material was available through purebreds, such as Yorkshires, Hampshires, and Landraces, raised by many small producers. Commercial breeding companies in the 1980s began developing different lines of pigs based on the genetics of the pure breeds in a system called crossbreeding. Modern swine crossbreeding techniques involve mating a boar (male) from a breed with rapid weight growth and sows (females) selected for their history of producing large litters.

Sows have a gestation period of 110–120 days with a 21-day interval between periods of estrus, the time during which they will accept mating by a boar. Sows have an average litter size of 12 piglets (somewhat fewer for a first pregnancy and somewhat more for certain Asian breeds), each piglet with a birth weight of about 1.4 kg (3 pounds), and typically produce two litters per year. A mature boar can mate as often as five to seven times per week. Gilts (young females) are usually mated by eight months of age and typically have a reproductive life of three to six litters, although individual sows may have 10 or more litters.

Most countries with developed pork production rely on artificial insemination. In fact, the semen from one boar ejaculate can be diluted to make 20 inseminations, each containing two to six billion sperm. In addition to reducing the number of boars needed for breeding, artificial insemination allows the selection of boars with the highest genetic merit, which results in more rapid improvement of the herd population. The semen may be collected and processed from boars raised by producers or purchased from stud farms that specialize in semen collection and marketing.

Piglets move to the sow's udder to begin nursing moments after birth and are weaned between two and five weeks, with about a 15 to 20 percent pre-weaning mortality rate from stillbirths and being crushed by the lactating sow. Pigs that weigh between about 18 and 57 kg (40 and 125 pounds) are known as growing pigs, from about 57 to 100 kg (125 to 220 pounds) as finishing pigs, and more than about 100 kg as hogs or market pigs because they are ready for butchering. Hogs are typically brought to market when they are five to six months old. Most males are castrated shortly after birth to avoid an off-flavour in their meat. Castrated males are called barrows.

Production Systems

Pork production can lend itself to mechanization and reduced use of high-priced labour. Self-feeders, diets composed of grains and oilseed by-products, and construction of slotted floors and outside tanks or lagoons for manure storage have become almost universal among large-scale commercial producers in developed countries. Particularly in developed countries, most pigs are raised indoors with various means of environmental control. Air-conditioned barns for excessively hot summers and heated floors and space heating or heat lamps for cold winters are widespread.

A heat lamp warming the litter of a
Yorkshire sow in a farrowing pen.

Production methods have evolved into systems divided by the stages of the pig's life cycle: birth, weaning, growth, finishing, and market. The three common operations are farrow-to-finish, farrow-to-feeder, and feeder-to-market. Farrowing refers to a sow giving birth. The farrow-to-finish operation is the historic foundation of the pork industry and includes all phases: breeding, gestation, farrowing, lactation, weaning, and subsequently growing the pigs to market weight. Typically, these operations have been on family farms, where owners raise pigs along with a grain operation in which much of the grain is fed to the pigs, saving the owner the cost of transporting and selling the grain. Additionally, the pig manure provides an excellent source of nitrogen, phosphorus, and potassium for fertilizing cropland. Historically, farrow-to-finish has been the most profitable type of hog enterprise. Many small-farm holders have full-time jobs in a nonfarming occupation and breed hogs to supplement their income.

Many pigs are now raised in vertically integrated systems, where ownership is maintained from the production farm through the meat-processing plant to the grocery store.

Farrow-to-feeder operations have the highest labour requirements, and many producers specialize in this part of the production cycle. It includes the management of the breeding herd, gestating sows, and piglets until they reach the growing (feeder) stage. The farmer retains control of the piglets until they are sold to another entity for feeder-to-market production. There are two common sale times—at early weaning, when a piglet weighs 5 to 7 kg (11 to 15 pounds), and at the start of the growing pig stage, when it weighs 18 to 25 kg (40 to 55 pounds) at about eight weeks. Most of these pigs are sold on a long-standing contract with a person involved in the final stage of production, feeder-to-market.

Feeder-to-market production has the lowest labour and management requirements. The producer in this stage purchases the feeder pigs and raises them to market weights in about 16 weeks. This part of the cycle requires the most feed and produces the most manure; therefore, it fits well with grain producers who have a lot of grain for feed and farmland that can use the pigs' manure as fertilizer. It is the least profitable per head, however, and two or three times as many pigs must be produced to earn as much as a farrow-to-finish producer.

Basic Dietary Requirements

Pigs have the same basic nutritional requirements as humans, which include water, various vitamins and minerals, protein for growth and repair, carbohydrates for energy, and fat to supply essential fatty acids that are not synthesized in adequate quantities. Water is often a forgotten nutrient because it is usually readily available. As a guide, pigs need two to three times as much water as dry feed, depending on environmental temperatures.

The fat-soluble vitamins that must be added to swine diets include vitamins A, D, E, and K. Water-soluble vitamins—in particular, the vitamin B complex—that must be added include niacin, pantothenic acid, riboflavin, and vitamin B12. Biotin, folic acid, and choline are sometimes recommended in diets of young pigs and the breeding herd. Vitamin requirements are usually listed as International Units, milligrams, or micrograms per unit of feed.

Mineral needs can be divided into major minerals and trace minerals. Major minerals that need to be added to the diet include calcium, phosphorus, and common salt. Requirements for major minerals are usually listed as a percentage of the diet. Trace minerals that need to be added to pig diets include iron, zinc, copper, manganese, iodine, and selenium. Although other minerals are required for growth, they are present in adequate amounts in feedstuffs. Requirements for trace minerals are usually listed as parts per million or milligrams per kilogram.

There is sufficient fat (about 1 percent) in the grain or feed of a pig's diet to supply all of its essential fatty acid requirements. Protein is a source of amino acids, 10 of which are deemed essential dietary requirements for pig nutrition. An additional 11 or so amino acids can be synthesized by the pig's metabolism and, although required for muscle growth, do not need to be present in the diet.

Corn (maize) is a favourite energy or carbohydrate source for pigs, but wheat, sorghum, milo, barley, and oats also are used if the price is favourable. Wherever abundant and reasonable in price, soybean oil meal is the favoured source of protein and amino acids, and other oil meals and high-protein by-products are used in most countries.

Special Dietary Requirements

The nutritional requirements of pigs vary according to their age, sex, and activities. For example, a boar's nutritional requirements are based on its weight and the number of times it has ejaculated, whether by inseminating sows or by having its semen collected for artificial insemination.

Nutrient requirements during gestation are much lower than would be expected; the major concern is that the sows do not become overweight before giving birth. Gilts should gain about 45 kg (100 pounds) during pregnancy. This weight gain includes about 14 kg (30 pounds) for offspring, another 14 kg for products of conception (increased weight of uterus and fluids), and 18 kg (40 pounds) of general weight gain. Sows, which have already produced litters, should gain 27 to 32 kg (60 to 70 pounds). A daily balanced diet of 1.8 kg (4 pounds) of feed will meet the nutritional requirements of gestating pigs in temperate environmental conditions.

After farrowing, a lactating sow's first milk is called colostrum, which lasts about three days. During this period, a sow needs 2 to 3 kg (4.5 to 6.5 pounds) of feed per day. Colostrum is very high in nutrients and factors that provide passive immunity to nursing piglets. This passive immunity is essential for disease resistance before piglets develop their own immunity, so all newborn piglets need to nurse immediately. Sows usually nurse their litters for two to five weeks, depending on the management system. Lactating sows have high nutrient requirements and at peak production may generate as much as 6 kg (13 pounds) of milk per day for their offspring. To prevent large weight losses in the sow, they need to be fed as much feed as they can consume. This can be as much as 10 to 12 kg (22 to 26 pounds) at three or more weeks after farrowing.

Weaned pigs are usually moved to a nursery where the temperature can be kept higher than 27 °C (80 °F) until they are about four weeks old. Piglets typically stay in the nursery for six to eight weeks. Newly weaned pigs have an immature digestive system, and their first diet after weaning until about four weeks of age should contain dried milk products in addition to energy and protein sources. Typically, nursery pigs are fed two to four different diets as they grow.

Growing pigs should be fed at least four distinct diets to optimize gain. As a pig grows, it eats more each day, but the nutrient density can be reduced.

Disease Prevention

The health of swine can best be ensured by a combination of prevention and treatment of diseases. Prevention includes both biosecurity and vaccination. Biosecurity includes isolating pigs from other species, both domestic and feral, as well as isolating pigs from each other by age. A major health risk is the introduction of new pigs into a resident population, because pigs brought from other farms are likely to carry disease-causing organisms to which the resident population has not developed any immunity. Human visitors also pose some risk, which can be mitigated by having them put on clean clothes and boots at a swine facility. A strict sanitation and traffic control program minimizes opportunities for new disease organisms to enter the herd, while systematic vaccination reduces the likelihood of routine diseases. A comprehensive herd health program also includes optimum nutrition, comfortable housing, excellent ventilation, and vigorous parasite control.

Safe and effective vaccines are available for many swine diseases, and producers work with their veterinarians to develop health programs that will alleviate infections of diseases prevalent in their local areas. Antibiotics may be added to the feed or water or be given by injection. Low-level doses of antibiotics, known as subtherapeutic, in the feed assist in preventing various bacteria from expressing disease symptoms. Infected pigs exhibiting disease symptoms may be treated with therapeutic levels. Producers treating pigs with any medication must be aware of and follow minimum withdrawal periods before the pigs are marketed.

Improvements in breeding, disease control, management, and feed formulation have all contributed to faster gains and lower feed requirements per kilogram of weight gain. The use of antibiotics began in the early 1950s in the United States, and the practice immediately resulted in increasing the rate of weight gain in nursery pigs (especially in regions with less favourable sanitation) by as much as 20 percent and by about 5 percent in pigs weighing more than 50 kg (110 pounds). Antibiotics became a standard ingredient in most young pigs' diets. Nevertheless, many European countries have restricted subtherapeutic use of antibiotics for growth promotion in livestock diets because of concern that antibiotic-resistant bacteria that infect humans may develop.

Common Diseases

Pigs are subject to many infectious and parasitic diseases. Diseases can be divided into infectious and noninfectious. Infectious diseases are transmitted between animals and include various bacterial, viral, and mycoplasmal organisms, as well as parasites. Noninfectious diseases include poisonous plants, toxins, nutritional excesses and deficiencies, and metabolic diseases such as ulcers.

Common diseases controllable by vaccination include transmissible gastroenteritis, which is often fatal to piglets (even when vaccinated); leptospirosis, which can also infect humans and most warm-blooded animals; pseudorabies, a viral disease that causes high mortality in piglets; and erysipelas, a bacterial infection that causes inflammation of the skin and swelling and stiffness of the joints. Cholera and foot-and-mouth disease, formerly controlled by vaccination, are now usually controlled by slaughter of infected herds. Necrotic enteritis and other infections of the intestinal tract are largely controlled by antibiotics. Atrophic rhinitis produces sneezing, crooked snouts, and poor performance and is controlled by a combination of vaccination and antibiotics.

Parasitic diseases can be divided into external and internal parasites. External parasites include lice and mites (which cause mange). Effective topical and internal preparations are available for their control or elimination. Internal parasites include various worms, which can be controlled through effective treatment with anthelmintics and through improvements in sanitation. Internal parasites are less of a problem when pigs are raised on slatted floors, which reduce spreading and re-infection by separating the pigs from their manure and other intermediary parasite hosts.

Common noninfectious diseases include mycotoxins (produced by molds and fungi present on various feedstuffs), ulcers, mange, and feeds accidentally contaminated by pesticides. Mycotoxins are best prevented by timely harvest of the grains and drying them to a moisture content that is not conducive to mold and fungal growth, usually 14 percent or less. Older nonpregnant pigs can be given lightly contaminated feed with minimal risk, whereas young pigs are more susceptible to mycotoxins.

Nutritional diseases are rare as a result of the availability of quality feedstuffs and excellent information regarding nutrient requirements. Nutrient deficiencies are usually the result of improper diet formulation over an extended time and occur most often in young, rapidly growing pigs. Nutrient excesses are not common, the major risk being that excesses of one or two nutrients may bind to other nutrients, thereby interfering with their efficient absorption in the digestive tract. Large nutrient excesses or deficiencies also may cause pigs to reduce their feed intake to prevent toxicity or nutrient disturbances.

Sheep

Sheep are able to subsist on sparse forage and limited water. Their wool is light in relation to its value and is relatively imperishable, both of which qualities enable wide exportation. During the

20th century, sheep-raising in some areas, particularly the western United States, has declined in favour of more profitable cattle.

Breeds

The gestation period for sheep is 147 days with 16.7 days between periods of estrus, which last 29 hours. The average number of lambs raised per hundred ewes is 91, and the average fleece weight per shearing is 8.34 pounds (3.78 kilograms).

Of more than 200 breeds of sheep in existence in the world, the majority are of limited interest except in the localities where they are raised. Sheep breeds are generally classified as medium wool, long wool, and fine wool. Of the medium wool breeds the Hampshire, Shropshire, Southdown, Suffolk, Oxford, and Dorset all originated in England. The Cheviot and Black Faced Highland originated in Scotland. The Panama, Columbia, and Targhee were developed in the United States, and the Corriedale in New Zealand. After World War II such larger breeds as the Suffolk and Hampshire increased in popularity at the expense of the smaller breeds.

The long wool breeds, including the Cotswold, Lincoln, Leicester, and Romney, were all developed in England and, in addition to mutton, produce wool of unusually long fibre length that is suitable for rugs and coarse fabrics.

The original fine-wool breed was the Merino, developed in Spain from stock native to that country before the Christian era. Though medieval Spain sought to preserve a monopoly on the Merino, the sheep gradually spread to France, Italy, and the rest of Europe. Today the Merino is prominent in Australia, the United States, Russia, South Africa, Argentina, France, and Germany; the breed is designated by various names such as Australian Merino in Australia and Merino Transhumante in Spain. The Merino was the main ancestor of the French Rambouillet, somewhat larger and less wrinkled than the Merino. This breed prospers in the western ranges of the United States, where two-thirds of that country's sheep are raised. The Corriedale breed, adapted to both farms and ranges, is especially valued in New Zealand and Australia. Most commercial sheep today represent two-breed or three-breed crosses, with white-faced crossbred ewes preferred in the range areas and a black-faced sire, such as Suffolk or Hampshire, preferred for market lambs, which are either finished for slaughter or sold as breeding ewes.

Feeding

Sheep are excellent foragers and, being ruminants, can utilize both pasture forage and harvested roughage. Selective in their grazing habits, they prefer short grass when available. Pregnant ewes can run on late pasture as long as it is available and abundant but in winter subsist satisfactorily on well-cured legume hay or mixed hay carrying a high percentage of legume. Corn (maize) silage is relatively inexpensive and relished by sheep; lactating ewes and lambs being finished for market usually require some concentrate, with corn (maize) favoured because of its high energy content and reasonable cost.

Range sheep grazing selectively on native plants frequently develop mild deficiencies of protein, energy, phosphorus, and vitamin A, especially when plants are mature or dormant or are eaten by ewes in the later stages of pregnancy or lactation. Broad spectrum antibiotics at the rate of five to 10 milligrams per pound of feed are normally used in all lamb finishing rations to prevent digestive disturbances and infections.

Diseases

Such internal parasites as the tapeworm and several species of roundworms that infest the gastrointestinal tract are perhaps the greatest scourge of sheep, but modern vermifuges are quite effective against these. Dips are used to combat such external parasites as ticks, lice, and mites. Foot rot, caused by an infection of the soft tissue between the toes, results in extreme lameness and even loss of the hoof. The more persistent type is caused by a specific organism that is difficult to treat. The pain and the restricted movement of infected sheep result in rapid loss of weight. Enterotoxemia, or pulpy kidney, affects lambs at two to six weeks of age, especially those starting on unusually lush or rich feeds. A vaccination is quite effective in preventing this otherwise costly ailment.

Goats

Probably first domesticated in the East, perhaps during prehistoric times, the goat has long been used as a source of milk, cheese, mohair, and meat. Its skin has been valued as a source for leather. In China, Great Britain, Europe, and North America, the goat is primarily a milk producer. By good management its limited (six months per year) breeding season and the consequent difficulty of maintaining a level supply of milk throughout the year, can be overcome. The goat is especially adapted to small-scale production of milk for the family table; one or two goats supply sufficient milk for a family throughout the year and can be maintained economically in quarters where it would not be practical to keep a cow.

Pure-white goat's milk compares favourably with cow's milk in flavour and keeping qualities under sanitary conditions. It has certain characteristics differing from cow's milk that make it more easily digested by infants, invalids, and persons with allergies. Goat flesh is edible, that from young kids being quite tender and more delicate in flavour than lamb, which it resembles. Goat flesh is much prized in the Mediterranean countries, particularly in Spain, Italy, the south of France, and Greece. The Angora and Cashmere breeds are famous for their fine wool or mohair.

The many breeds may be roughly grouped: the prickeared—e.g., Swiss goats; the eastern, or Nubian, with long, drooping ears; and the wool goat—e.g., Angora. While it is usually easy to distinguish goats from sheep, certain hair breeds of the latter are, to the layman, only distinguishable from goats by the direction of the tail, upward in goats, downward in sheep.

Of the Swiss goats, from which many of the best modern breeds are derived, the Toggenburg and Saanen are most important. The French breeds have much Swiss blood. In Germany the many varieties trace to Swiss breeds, which are also popular throughout Scandinavia and the Netherlands.

The Maltese goat, an important source of milk on the island of Malta, probably contains eastern blood. Many goats are found in Spain, northern Africa, and Italy, among them the Murcian, Granada, and La Mancha.

Nubians are African goats, chiefly Egyptian. They are usually large, short-haired goats with large lop ears and Roman noses. They may be of solid colour, parti-coloured, or spotted. The goats in Israel and Syria have long hair and large lop ears and most commonly are solid black or with white spots.

In Britain, the native goat was small, with short legs, long hair—usually gray but of no fixed colour—and with no definite markings. The widespread use of pedigree males, mostly of Swiss extraction, to improve the milk yield, has resulted in the almost total disappearance of the native types.

Horses

Horses were among the last species of livestock to be domesticated. Domestication took place at least as early as 3000 BCE, probably in the Near East. The wild ass, which when domesticated is usually called a donkey, was first domesticated in Egypt about 3400 BCE.

Breeds

The Arabian, the oldest recognized breed of horse in the world, is thought to have originated in Arabia before 600 CE. Though its history is lost in the past, the breed probably descended from the Libyan horse, which in turn was probably preceded by horses of similar characteristics in Assyria, Greece, and Egypt as early as 1000 BCE. The Arabian may be bay, gray, chestnut, brown, black, or white in hair colour but always has a black skin. It ranges from 14.1 to 15.1 hands (4.7 to 5.0 feet, or 1.4 to 1.5 metres) in height. The Arabian horse has one lumbar vertebra less than other breeds of horse and is characterized by the high carriage of its head, long neck, and spirited action.

The Thoroughbred racing horse is descended from three desert stallions brought to England between 1689 and 1724; all of the Thoroughbreds of the world today trace their ancestry to one of these stallions.

The American Saddle Horse, which originated in the United States, was formed by crossing Thoroughbreds, Morgans, and Standardbreds on native mares possessing an easy gait. The American Saddle Horse is 15 to 16 hands (5 to 5.3 feet, or 1.5 to 1.6 metres) in height. Its colours are bay, brown, black, gray, and chestnut. There are two distinct types of the American Saddle Horse: three-gaited and five-gaited. The three natural gaits are walk, trot, and canter. Three-gaited saddle horses are shown with a short tail and cropped mane. They often have slightly less style and finish than the five-gaited horse. The five-gaited saddle horse has the three natural gaits plus the rack and a slow gait, which is usually a stepping pace. The American Saddle Horse is also used as a fine harness horse mainly for show.

The American Quarter Horse traces to the Thoroughbred, and includes the blood of other breeds, such as the Morgan, the American Saddle Horse, and several strains of native horses. This fast, muscular horse has been raced, ridden in rodeos, and used for herding cattle.

The typical Quarter Horse is 15 to 16 hands tall and is of powerful build, suitable for both racing and the rough life of a cow pony. This horse is noted for its intelligence, easy disposition, and cow sense.

The Tennessee Walking Horse, or plantation horse, traces mainly to the Standardbred but also includes Thoroughbred and American Saddle Horse blood. The Tennessee Walking Horse is noted for its running walk, a slowgliding gait in which the hind foot oversteps the print of the front foot by as much as 24 inches (600 millimetres). This breed is 15.2 to 16 hands high and is bay, black, chestnut, roan, or gray in colour.

The Morgan traces directly to "the Justin Morgan horse," foaled in 1793, of unknown breeding but no doubt tracing to Arabian stock. A dark bay in colour, Morgan stood 14 hands high and weighed 950 pounds (430 kilograms). He was a heavily muscled, short-legged horse of great style, quality, and endurance. He is the world's best example of prepotency, since he alone founded the Morgan breed. The Morgan is used for both riding and driving. It ranges from 14 to 16 hands in height and resembles the Arabian in size, conformation, quality, and endurance.

The American Standardbred originated around New York City during the first half of the 19th century from Thoroughbred, Morgan, Norfolk Trotter, Arabian, Barb, and pacers of mixed breeding. The modern Standardbred is smaller than the Thoroughbred, ranging from 15 to 16 hands in height and averaging about 15.2 hands. In racing condition it weighs from 900 to 1,000 pounds (410–450 kilograms). Stallions in stud condition average from 1,100 to 1,200 pounds (500–545 kilograms). Compared with the Thoroughbred, the Standardbred is longer-bodied, shorter-legged, heavier-boned, and stockier in build. Prevailing colours are bay, brown, and chestnut.

Draft horses have largely been supplanted by trucks and tractors in the developed countries of the world. Major draft breeds include the Percheron, developed in France; the Clydesdale of Scotland; the Shire of England; the Suffolk of England; and the Belgian of Belgium. These breeds range from $15^{1}/_{2}$ to 17 hands in height at the withers; at maturity the mares weigh from 1,600 to 2,000 pounds (720–900 kilograms) and the stallions from 1,900 to 2,200 pounds (860–1000 kilograms).

The more popular pony breeds are the Shetland, which originated in the Shetland Islands, and the Hackney, of English origin. Ponies must be under 14.2 hands in height at the withers and are used both for show and for children's pleasure.

Selected breeds of ponies						
	Name	Origin	Height (hands)	Aptitude	Characteristics	Comments
	Connemara	Ireland	13–14.2	Riding; light draft	Well-formed hindquarters with high-set tail; long neck with full mane; well-muscled legs.	Ireland's only indigenous breed; extremely hardy; known for its exceptional jumping ability and the ease of its gait.
	Pony of the americas	U.S.	11.2–13.2	Riding	Appaloosa colouring; well-pricked ears; large, prominent eyes.	Cross between a shetland pony stallion and an appaloosa mare; developed as a versatile child's mount.
	Shetland	Shetland islands, scotland	10	Riding, light draft	Thick mane and tail; small head with pronounced jaw; short, muscular neck.	Thought to have existed since the bronze age; very powerful; used as a pit pony in mines of great britain in the 19th century; a popular child's mount.
	Welsh	Wales	12.2–13.2	Riding, light draft	Fine head with large eyes and small ears; typically gray in colour.	Very hardy; arabian influence; excellent gaits.

Feeding

The specific and exact nutrient requirements of horses are poorly understood. Usually, these may be supplied economically from pasture forage, harvested roughages, and concentrates. Good quality grass-legume pastures, in addition to iodized or trace-mineralized salt, will supply adequate nutrients to maintain an adult horse at light work (such as pulling a small cart) or mares during pregnancy. Lush, early spring pasture is very high in water and protein contents and may need to be supplemented with a high-energy source, such as grain, to meet the needs of horses performing medium to heavy work (such as plowing). Conversely, late fall- and winter-pasture forage is low in water and protein and may require protein and vitamin A supplementation. High-quality legume hays, such as early bloom alfalfa, are preferred for horses, especially those that are growing or lactating. Moldy or dusty feeds should be avoided because horses are extremely susceptible to forage poisoning and respiratory complications. Grass hays, such as timothy, prairie grass, orchard grass, and bluegrass, were preferred by early horsemen, especially for race horses, because they were usually free from mold and dust and tended to slow down the rate of passage through the intestinal tract. These hays are low in digestible energy and protein, however, and must be adequately supplemented. Silages of all sorts should be avoided since horses and mules are extremely susceptible to botulism and digestive upsets.

Oats are the preferred grain for horses because of their bulk. Corn (maize), barley, wheat, and milo can be used, however, whenever they are less expensive. Weanling foals require three pounds of feed per hundred pounds of live weight per day; as they approach maturity, this requirement drops to one pound of feed per hundred pounds of live weight daily. Horses normally reach mature weight at less than four years of age and 80 percent of their mature weight at less than two years of age.

A large and ever-growing number of horses stabled in cities and suburbs where sufficient roughages cannot be grown provide a large market for complete horse rations, including roughage, which are tailored to the total needs of specific animals according to their particular function at a given time, such as growth, pregnancy, lactation, or maintenance.

Horses will vary from the normal requirement in terms of weight, temperament, and previous nutrition. Foals will eat some pasture grass, forage, or hay when they are three days old and grain when they are three weeks old.

Diseases

Horses are especially susceptible to tetanus or lockjaw but can be given two-year protection through the use of a commonly accepted toxoid. There are two common types of abortion in horses: virus abortion, specifically viral rhinopneumonitis, and the Salmonella type. The former, which produces an influenza with pinkeye, catarrh, general illness, and abortion, affects both mares and foals, but all surviving horses develop natural resistance soon after infection. Pregnant mares thought to be subjected to infection may be given some protection by available vaccines. The Salmonella type of abortion can be prevented completely by vaccination. Encephalomyelitis, or sleeping sickness, is prevented by vaccination. A specific vaccine is available for anthrax, which is prevalent in Asia. Hemolytic anemia of foals has become a problem. Foals so afflicted are born normal but soon become sluggish and progressively weaker; the membranes of their eyes, mouth, and lips become very pale and the heartbeat becomes rapid. This condition is caused by antibodies in the mare's milk that destroy the foal's red blood cells. These antibodies are caused by the difference in blood type between the foal and the mother. Newborn foals can be muzzled to avoid nursing while their blood is checked for reaction against the serum and milk of its mother. Where reactions are noted, the mare is hand-milked at hourly intervals for 12 to 24 hours, and the foal is fed milk from another suitable mare or a milk substitute. Horses are quite susceptible to various infections, but rotation of pastures, strict sanitation, and the use of suitable vermifuges are quite effective.

Donkeys and Mules

Donkey (Equus asinus), a domesticated ass.

The words donkey and ass are generally used interchangeably to denote the same animal, though ass is more properly employed when the animal is wild (e.g., Equus africanus or E. hemionus) and donkey for a domesticated beast (E. asinus). Wild asses inhabit arid semidesert plains in Africa and Asia where the vegetation is sparse and coarse; the domestic donkey does well on coarse food and is hardy under rough conditions, hence its usefulness to humanity as a beast of burden in places where horses cannot flourish, such as the mountains of Ethiopia and other parts of northeast Africa, the high plains of Tibet, and the arid regions of Mongolia.

The donkey's occasional obstinacy in refusing work too heavy for it has become proverbial, but its equally proverbial stupidity is often a reaction to brutal treatment and neglect. It is naturally patient and persevering, responding to gentle treatment with affection and attachment to its master. One of the largest donkey breeds, the Mammoth Jack, was developed in the United States in the late 18th century from European imports, including the Adalusian, the Maltese, the Majorcan, the Poitou, and various Italian strains. It stands 15 to 16 hands (1.5 to 1.6 metres, or 4.9 to 5.2 feet) in height and weighs 410–520 kilograms (900–1,150 pounds) at maturity. The development of the breed was originally undertaken by George Washington and Henry Clay, among others, to produce larger, stronger mules for American industry.

The mule is produced by crossing a jackass (e.g., male donkey) with a mare. At one time many different types of mules were recognized, such as draft mules, farm mules, sugar mules, cotton mules, and mining mules, in declining order of size. The mining mule, a small rugged individual weighing as little as 270 kilograms (600 pounds), was used in pit mines. Mules are still used in some of the subtropical and tropical countries because of their ability to withstand most types of stress including heat, irregular feeding, and abuse. Mules are surer-footed than horses and are considered to be more intelligent. For that reason they are still used as saddle and pack mounts in precarious terrain. Unlike horses, mules refuse to damage themselves by overeating or by thrashing around when tangled up or in cramped quarters. The reverse cross of a stallion on a jenny (e.g., female ass) is called a hinny, which is slightly smaller than a mule; both mules and hinnies are sterile.

Two mules grazing in a snowy pasture in Idaho. Mules are formed by crossing a male donkey with a mare and cannot reproduce.

Buffalo

The name buffalo is applied to several different cud-chewing (ruminant) mammals of the ox family (Bovidae). The true, or Indian, buffalo (Bubalus bubalis), also known as water buffalo, or arna, exists both as a wild and domestic animal; it has been domesticated in Asia from very early times and was introduced into Italy about the year 600. A large ox-like animal of massive and rather clumsy build with large horns that are triangular in cross section, the Indian buffalo, standing five feet (1.5 metres) at the shoulder, has a dull black body, often very sparsely covered with hair. The horns, which may be over six feet (1.8 metres) long, spread outward and upward, approaching each other toward the tips; they meet more or less in one plane above the rounded forehead and elongated face. Used for draft purposes, and also for milk and butter, the domesticated Indian buffalo is found throughout the warmer parts of the Old World from China to Egypt, and in Hungary, France, and Italy. Its cousin, the Cape, or African, buffalo (Syncerus caffer), a black animal of similarly massive build, has never been domesticated.

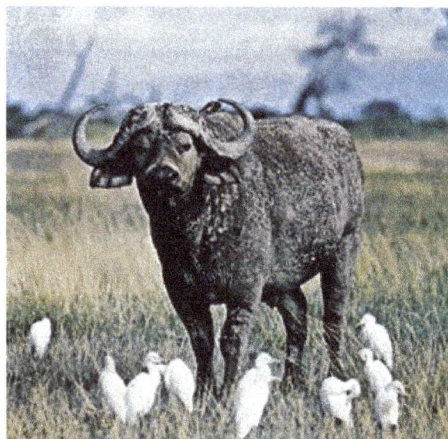

Cape or African buffalo (Syncerus caffer).

Camels

The term camel usually applies to two species of the genus Camelus. The Arabian camel, Camelus dromedarius, has one hump, the Bactrian camel, Camelus bactrianus, has two. The limbs are long and the feet have no traces of the second or fifth toes; the wide-spreading soft feet are well adapted for walking upon sand or snow. Horny pads on the chest and knees support the camel's weight when kneeling.

Dromedary or Arabian camel (Camelus dromedarius).

The Bactrian camel occurs throughout the highlands of Central Asia from Turkistan to Mongolia and is an important beast of burden throughout that region. The Arabian camel, characteristic of India, the Near East, and North Africa, is likewise primarily important as a beast of burden, though it also provides wool, milk, hides, and meat. It is longer-legged, shorter coated, and more lightly built than the Bactrian camel, standing about seven feet (2.1 metres) tall at the shoulder. In the 19th century camels were introduced to the U.S.–Mexico border regions, the Pacific Northwest, and Australia. The North American experiments were short-lived, but the animals were used in the exploration and development of the Australian outback until about 1940.

Camels can flourish on the coarsest of sparse vegetation, feeding on thorny plants, the leaves and twigs of shrubs, and dried grasses that other animals would refuse, though camels are not averse to more attractive food if it is available. When the feeding is good they accumulate stores of fat in their humps, upon which they are able to draw when conditions are adverse not only for sustenance but also for the manufacture of water by the oxidation of the fat; but they do not store water in the miscalled water cells. They are thus able to fast and go without drinking for several days; they have been known to go without water for 17 days and survive. Other adaptations that enable them to survive in deserts and other unfavourable environments include double rows of heavy protective eyelashes, haired ear openings, the ability to close their nostrils, and keen senses of sight and smell. The female produces one young at a birth after a gestation of 11 months and suckles it for a year; maturity is reached at the age of 10 to 12 years, and the life span is 30 to 40 years.

INTENSIVE ANIMAL FARMING

Intensive animal farming or industrial livestock production, also known by its opponents as factory farming, is a type of intensive agriculture, specifically an approach to animal husbandry designed to maximize production, while minimizing costs. To achieve this, agribusinesses keep livestock such as cattle, poultry, and fish at high stocking densities, at large scale, and using modern machinery, biotechnology, and global trade. The main products of this industry are meat, milk and eggs for human consumption. There are issues regarding whether factory farming is sustainable or ethical.

There is a continuing debate over the benefits, risks and ethics of intensive animal farming. The issues include the efficiency of food production; animal welfare; health risks and the environmental impact (e.g. agricultural pollution and climate change).

Types

Intensive farms hold large numbers of animals, typically cows, pigs, turkeys, or chickens, often indoors, typically at high densities. The aim is to produce large quantities of meat, eggs, or milk at the lowest possible cost. Food is supplied in place. Methods employed to maintain health and improve production may include the use of disinfectants, antimicrobial agents, anthelmintics, hormones and vaccines; protein, mineral and vitamin supplements; frequent health inspections; biosecurity; and climate-controlled facilities. Physical restraints, e.g. fences or creeps, are used to control movement or actions regarded as undesirable. Breeding programs are used to produce animals more suited to the confined conditions and able to provide a consistent food product.

Intensive production of livestock and poultry is widespread in developed nations. For 2002-2003, FAO estimates of industrial production as a percentage of global production were 7 percent for beef and veal, 0.8 percent for sheep and goat meat, 42 percent for pork, and 67 percent for poultry meat. Industrial production was estimated to account for 39 percent of the sum of global production of these meats and 50 percent of total egg production. In the U.S., according to its National Pork Producers Council, 80 million of its 95 million pigs slaughtered each year are reared in industrial settings.

Chickens

Hens in Brazil

The major milestone in 20th century poultry production was the discovery of vitamin D, which made it possible to keep chickens in confinement year-round. Before this, chickens did not thrive during the winter (due to lack of sunlight), and egg production, incubation, and meat production in the off-season were all very difficult, making poultry a seasonal and expensive proposition. Year-round production lowered costs, especially for broilers.

At the same time, egg production was increased by scientific breeding. After a few false starts, (such as the Maine Experiment Station's failure at improving egg production) success was shown by Professor Dryden at the Oregon Experiment Station.

Improvements in production and quality were accompanied by lower labor requirements. In the 1930s through the early 1950s, 1,500 hens provided a full-time job for a farm family in America. In the late 1950s, egg prices had fallen so dramatically that farmers typically tripled the number of hens they kept, putting three hens into what had been a single-bird cage or converting their floor-confinement houses from a single deck of roosts to triple-decker roosts. Not long after this, prices fell still further and large numbers of egg farmers left the business. This fall in profitability was accompanied by a general fall in prices to the consumer, allowing poultry and eggs to lose their status as luxury foods.

Robert Plamondon reports that the last family chicken farm in his part of Oregon, Rex Farms, had 30,000 layers and survived into the 1990s. However, the standard laying house of the current operators is around 125,000 hens.

The vertical integration of the egg and poultry industries was a late development, occurring after all the major technological changes had been in place for years (including the development of modern broiler rearing techniques, the adoption of the Cornish Cross broiler, the use of laying cages).

By the late 1950s, poultry production had changed dramatically. Large farms and packing plants could grow birds by the tens of thousands. Chickens could be sent to slaughterhouses for butchering and processing into prepackaged commercial products to be frozen or shipped fresh to markets or wholesalers. Meat-type chickens currently grow to market weight in six to seven weeks, whereas only fifty years ago it took three times as long. This is due to genetic selection and nutritional modifications (but not the use of growth hormones, which are illegal for use in poultry in the US and many other countries). Once a meat consumed only occasionally, the common availability and lower cost has made chicken a common meat product within developed nations. Growing concerns over the cholesterol content of red meat in the 1980s and 1990s further resulted in increased consumption of chicken.

Today, eggs are produced on large egg ranches on which environmental parameters are well controlled. Chickens are exposed to artificial light cycles to stimulate egg production year-round. In addition, forced molting is commonly practiced, in which manipulation of light and food access triggers molting, with the goal of increased egg size and production. Forced molting is controversial. While it is widespread in the US, it is prohibited in the EU.

On average, a chicken lays one egg a day, but not on every day of the year. This varies with the breed and time of year. In 1900, average egg production was 83 eggs per hen per year. In 2000, it was well over 300. In the United States, laying hens are butchered after their second egg laying season. In Europe, they are generally butchered after a single season. The laying period begins when the hen is about 18–20 weeks old (depending on breed and season). Males of the egg-type breeds have little commercial value at any age, and all those not used for breeding (roughly fifty percent of all egg-type chickens) are killed soon after hatching. The old hens also have little commercial value. Thus, the main sources of poultry meat 100 years ago (spring chickens and stewing hens) have both been entirely supplanted by meat-type broiler chickens.

Pigs

Pigs confined to a barn in an intensive
system, Midwestern United States.

Intensive piggeries (or hog lots) are a type of what in America is called a Concentrated Animal Feeding Operation (CAFO), specialized for the raising of domestic pigs up to slaughter weight. In this system, grower pigs are housed indoors in group-housing or straw-lined sheds, whilst pregnant sows are confined in sow stalls (gestation crates) and give birth in farrowing crates.

The use of sow stalls has resulted in lower production costs and concomitant animal welfare concerns. Many of the world's largest producers of pigs (U.S. and Canada) use sow stalls, but some nations (e.g. the UK) and US States (such as Florida and Arizona) have banned them.

Intensive piggeries are generally large warehouse-like buildings. Indoor pig systems allow the pig's condition to be monitored, ensuring minimum fatalities and increased productivity. Buildings are ventilated and their temperature regulated. Most domestic pig varieties are susceptible to heat stress, and all pigs lack sweat glands and cannot cool themselves. Pigs have a limited tolerance to high temperatures and heat stress can lead to death. Maintaining a more specific temperature within the pig-tolerance range also maximizes growth and growth to feed ratio. In an intensive operation pigs will lack access to a wallow (mud), which is their natural cooling mechanism. Intensive piggeries control temperature through ventilation or drip water systems (dropping water to cool the system).

Pigs are naturally omnivorous and are generally fed a combination of grains and protein sources (soybeans, or meat and bone meal). Larger intensive pig farms may be surrounded by farmland where feed-grain crops are grown. Alternatively, piggeries are reliant on the grains industry. Pig feed may be bought packaged or mixed on-site. The intensive piggery system, where pigs are confined in individual stalls, allows each pig to be allotted a portion of feed. The individual feeding system also facilitates individual medication of pigs through feed. This has more significance to intensive farming methods, as the close proximity to other animals enables diseases to spread more rapidly. To prevent disease spreading and encourage growth, drug programs such as antibiotics, vitamins, hormones and other supplements are preemptively administered.

Indoor systems, especially stalls and pens (i.e. 'dry,' not straw-lined systems) allow for the easy collection of waste. In an indoor intensive pig farm, manure can be managed through a lagoon system or other waste-management system. However, odor remains a problem which is difficult to manage.

The way animals are housed in intensive systems varies. Breeding sows spend the bulk of their time in sow stalls during pregnancy or farrowing crates, with their litters, until market.

Piglets often receive range of treatments including castration, tail docking to reduce tail biting, teeth clipped (to reduce injuring their mother's nipples and prevent later tusk growth) and their ears notched to assist identification. Treatments are usually made without pain killers. Weak runts may be slain shortly after birth.

Piglets also may be weaned and removed from the sows at between two and five weeks old and placed in sheds. However, grower pigs - which comprise the bulk of the herd - are usually housed in alternative indoor housing, such as batch pens. During pregnancy, the use of a stall may be preferred as it facilitates feed-management and growth control. It also prevents pig aggression (e.g. tail biting, ear biting, vulva biting, food stealing). Group pens generally require higher stockmanship skills. Such pens will usually not contain straw or other material. Alternatively, a straw-lined shed may house a larger group (i.e. not batched) in age groups.

Cattle

Cattle are domesticated ungulates, a member of the family Bovidae, in the subfamily Bovinae, and descended from the aurochs (Bos primigenius). They are raised as livestock for meat (called beef and veal), dairy products (milk), leather and as draught animals. As of 2009–2010 it is estimated that there are 1.3–1.4 billion head of cattle in the world.

Beef cattle on a feedlot in the Texas Panhandle. Such confinement creates more work for the farmer but allows the animals to grow rapidly.

The most common interactions with cattle involve daily feeding, cleaning and milking. Many routine husbandry practices involve ear tagging, dehorning, loading, medical operations, vaccinations and hoof care, as well as training for agricultural shows and preparations.

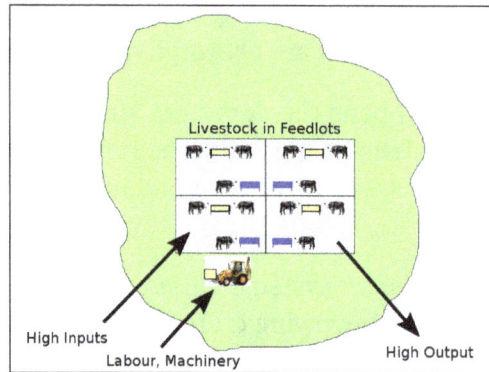

Diagram of feedlot system. This can be contrasted
with more traditional grazing systems.

Once cattle obtain an entry-level weight, about 650 pounds (290 kg), they are transferred from the range to a feedlot to be fed a specialized animal feed which consists of corn byproducts (derived from ethanol production), barley, and other grains as well as alfalfa and cottonseed meal. The feed also contains premixes composed of microingredients such as vitamins, minerals, chemical preservatives, antibiotics, fermentation products, and other essential ingredients that are purchased from premix companies, usually in sacked form, for blending into commercial rations. Because of the availability of these products, a farmer using their own grain can formulate their own rations and be assured the animals are getting the recommended levels of minerals and vitamins.

There are many potential impacts on human health due to the modern cattle industrial agriculture system. There are concerns surrounding the antibiotics and growth hormones used, increased *E. coli* contamination, higher saturated fat contents in the meat because of the feed, and also environmental concerns.

As of 2010, in the U.S. 766,350 producers participate in raising beef. The beef industry is segmented with the bulk of the producers participating in raising beef calves. Beef calves are generally raised in small herds, with over 90% of the herds having less than 100 head of cattle. Fewer producers participate in the finishing phase which often occurs in a feedlot, but nonetheless there are 82,170 feedlots in the United States.

Controversies and Criticisms

Advocates of factory farming claim that factory farming has led to the betterment of housing, nutrition, and disease control over the last twenty years, while opponents claim that it harms wildlife and the environment, creates health risks, abuses animals, and raises ethical issues.

In the UK, the Farm Animal Welfare Council was set up by the government to act as an independent advisor on animal welfare in 1979 and expresses its policy as five freedoms: from hunger & thirst; from discomfort; from pain, injury or disease; to express normal behavior; from fear and distress.

There are differences around the world as to which practices are accepted and there continue to be changes in regulations with animal welfare being a strong driver for increased regulation. For example, the EU is bringing in further regulation to set maximum stocking densities for meat chickens by 2010, where the UK Animal Welfare Minister commented, "The welfare of meat chickens is a major concern to people throughout the European Union. This agreement sends a strong message to the rest of the world that we care about animal welfare."

Factory farming is greatly debated throughout Australia, with many people disagreeing with the methods and ways in which the animals in factory farms are treated. Animals are often under stress from being kept in confined spaces and will attack each other. In an effort to prevent injury leading to infection, their beaks, tails and teeth are removed. Many piglets will die of shock after having their teeth and tails removed, because painkilling medicines are not used in these operations. Factory farms are a popular way to gain space, with animals such as chickens being kept in spaces smaller than an A4 page.

For example, in the UK, de-beaking of chickens is deprecated, but it is recognized that it is a method of last resort, seen as better than allowing vicious fighting and ultimately cannibalism. Between 60 and 70 percent of six million breeding sows in the U.S. are confined during pregnancy, and for most of their adult lives, in 2 by 7 ft (0.61 by 2.13 m) gestation crates. According to pork producers and many veterinarians, sows will fight if housed in pens. The largest pork producer in the U.S. said in January 2007 that it will phase out gestation crates by 2017. They are being phased out in the European Union, with a ban effective in 2013 after the fourth week of pregnancy. With the evolution of factory farming, there has been a growing awareness of the issues amongst the wider public, not least due to the efforts of animal rights and welfare campaigners. As a result, gestation crates, one of the more contentious practices, are the subject of laws in the U.S., Europe and around the world to phase out their use as a result of pressure to adopt less confined practices.

Death rates for sows have been increasing in the US from prolapse, which has been attributed to intensive breeding practices. Sows produce on average 23 piglets a year.

Human Health Impact

According to the U.S. Centers for Disease Control and Prevention (CDC), farms on which animals are intensively reared can cause adverse health reactions in farm workers. Workers may develop acute and chronic lung disease, musculoskeletal injuries, and may catch infections that transmit from animals to human beings (such as tuberculosis).

Pesticides are used to control organisms which are considered harmful and they save farmers money by preventing product losses to pests. In the US, about a quarter of pesticides used are used in houses, yards, parks, golf courses, and swimming pools and about 70% are used in agriculture. However, pesticides can make their way into consumers' bodies which can cause health problems. One source of this is bioaccumulation in animals raised on factory farms.

Studies have discovered an increase in respiratory, neurobehavioral, and mental illnesses among the residents of communities next to factory farms.

The CDC writes that chemical, bacterial, and viral compounds from animal waste may travel in the soil and water. Residents near such farms report problems such as unpleasant smell, flies and adverse health effects.

The CDC has identified a number of pollutants associated with the discharge of animal waste into rivers and lakes, and into the air. Antibiotic use in livestock may create antibiotic-resistant pathogens; parasites, bacteria, and viruses may be spread; ammonia, nitrogen, and phosphorus can reduce oxygen in surface waters and contaminate drinking water; pesticides and hormones may cause hormone-related changes in fish; animal feed and feathers may stunt the growth of desirable plants in surface waters and provide nutrients to disease-causing micro-organisms; trace elements such as arsenic and copper, which are harmful to human health, may contaminate surface waters.

Intensive farming may make the evolution and spread of harmful diseases easier. Many communicable animal diseases spread rapidly through densely spaced populations of animals and crowding makes genetic reassortment more likely. However, small family farms are more likely to introduce bird diseases and more frequent association with people into the mix, as happened in the 2009 flu pandemic.

In the European Union, growth hormones are banned on the basis that there is no way of determining a safe level. The UK has stated that in the event of the EU raising the ban at some future date, to comply with a precautionary approach, it would only consider the introduction of specific hormones, proven on a case by case basis. In 1998, the European Union banned feeding animals antibiotics that were found to be valuable for human health. Furthermore, in 2006 the European Union banned all drugs for livestock that were used for growth promotion purposes. As a result of these bans, the levels of antibiotic resistance in animal products and within the human population showed a decrease. The international trade in animal products increases the risk of global transmission of virulent diseases such as swine fever, BSE, foot and mouth and bird flu.

In the United States, the use of antibiotics in livestock is still prevalent. The FDA reports that 80 percent of all antibiotics sold in 2009 were administered to livestock animals, and that many of these antibiotics are identical or closely related to drugs used for treating illnesses in humans. Consequently, many of these drugs are losing their effectiveness on humans, and the total healthcare costs associated with drug-resistant bacterial infections in the United States are between $16.6 billion and $26 billion annually.

Methicillin-resistant Staphylococcus aureus (MRSA) has been identified in pigs and humans raising concerns about the role of pigs as reservoirs of MRSA for human infection. One study found that 20% of pig farmers in the United States and Canada in 2007 harbored MRSA. A second study revealed that 81% of Dutch pig farms had pigs with MRSA and 39% of animals at slaughter carried the bug were all of the infections were resistant to tetracycline and many were resistant to other antimicrobials. A more recent study found that MRSA ST398 isolates were less susceptible to tiamulin, an antimicrobial used in agriculture, than other MRSA or methicillin susceptible *S. aureus*. Cases of MRSA have increased in livestock animals. CC398 is a new clone of MRSA that has emerged in animals and is found in intensively reared production animals (primarily pigs, but also cattle and poultry), where it can be transmitted to humans. Although dangerous to humans, CC398 is often asymptomatic in food-producing animals.

A 2011 nationwide study reported nearly half of the meat and poultry sold in U.S. grocery stores — 47 percent — was contaminated with S. aureus, and more than half of those bacteria — 52 percent — were resistant to at least three classes of antibiotics. Although Staph should be killed with proper cooking, it may still pose a risk to consumers through improper food handling and cross-contamination in the kitchen. The fact that drug-resistant S. aureus was so prevalent, and likely came from the food animals themselves, is troubling, and demands attention to how antibiotics are used in food-animal production today.

In April 2009, lawmakers in the Mexican state of Veracruz accused large-scale hog and poultry operations of being breeding grounds of a pandemic swine flu, although they did not present scientific evidence to support their claim. A swine flu which quickly killed more than 100 infected persons in that area, appears to have begun in the vicinity of a Smithfield subsidiary pig CAFO (concentrated animal feeding operation).

Environmental Impact

Intensive factory farming has grown to become the biggest threat to the global environment through the loss of ecosystem services and global warming. It is a major driver to global environmental degradation and biodiversity loss. The process in which feed needs to be grown for animal use only is often grown using intensive methods which involve a significant amount of fertiliser and pesticides. This sometimes results in the pollution of water, soil and air by agrochemicals and manure waste, and use of limited resources such as water and energy at unsustainable rates. Entomophagy is evaluated by many experts as a sustainable solution to traditional livestock, and, if intensively farmed on a large-scale, would cause a far-lesser amount of environmental damage.

Industrial production of pigs and poultry is an important source of Greenhouse gas emissions and is predicted to become more so. On intensive pig farms, the animals are generally kept on concrete with slats or grates for the manure to drain through. The manure is usually stored in slurry form (slurry is a liquid mixture of urine and feces). During storage on farm, slurry emits methane and when manure is spread on fields it emits nitrous oxide and causes nitrogen pollution of land and water. Poultry manure from factory farms emits high levels of nitrous oxide and ammonia.

Large quantities and concentrations of waste are produced. Air quality and groundwater are at risk when animal waste is improperly recycled.

Environmental impacts of factory farming include:

- Deforestation for animal feed production.
- Unsustainable pressure on land for production of high-protein/high-energy animal feed.
- Pesticide, herbicide and fertilizer manufacture and use for feed production.
- Unsustainable use of water for feed-crops, including groundwater extraction.
- Pollution of soil, water and air by nitrogen and phosphorus from fertiliser used for feed-crops and from manure.
- Land degradation (reduced fertility, soil compaction, increased salinity, desertification).
- Loss of biodiversity due to eutrophication, acidification, pesticides and herbicides.
- Worldwide reduction of genetic diversity of livestock and loss of traditional breeds.
- Species extinctions due to livestock-related habitat destruction (especially feed-cropping).

Labor

Small farmers are often absorbed into factory farm operations, acting as contract growers for the industrial facilities. In the case of poultry contract growers, farmers are required to make costly investments in construction of sheds to house the birds, buy required feed and drugs - often settling for slim profit margins, or even losses.

Research has shown that many immigrant workers in concentrated animal farming operations (CAFOs) in the United States receive little to no job-specific training or safety and health information regarding the hazards associated with these jobs. Workers with limited English proficiency are significantly less likely to receive any work-related training, since it is often only provided in English. As a result, many workers do not perceive their jobs as dangerous. This causes inconsistent personal protective equipment (PPE) use, and can lead to workplace accidents and injuries. Immigrant workers are also less likely to report any workplace hazards and injuries.

Market Concentration

The major concentration of the industry occurs at the slaughter and meat processing phase, with only four companies slaughtering and processing 81 percent of cows, 73 percent of sheep, 57 percent of pigs and 50 percent of chickens. This concentration at the slaughter phase may be in large part due to regulatory barriers that may make it financially difficult for small slaughter plants to be built, maintained or remain in business. Factory farming may be no more beneficial to livestock producers than traditional farming because it appears to contribute to overproduction that drives down prices. Through "forward contracts" and "marketing agreements", meatpackers are able to set the price of livestock long before they are ready for production. These strategies often cause farmers to lose money, as half of all U.S. family farming operations did in 2007.

In 1967, there were one million pig farms in America; as of 2002, there were 114,000. Many of the nation's livestock producers would like to market livestock directly to consumers but with limited USDA inspected slaughter facilities, livestock grown locally can not typically be slaughtered and processed locally.

ANIMAL FEED

A feedlot in Texas, USA, where cattle are
"finished" (fattened on grains) prior to slaughter.

Animal feed is food given to domestic animals in the course of animal husbandry. There are two basic types: *fodder* and *forage*. Used alone, the word "feed" more often refers to fodder.

Fodder

Equine nutritionists recommend that 50% or more of
a horse's diet by weight should be forages, such as hay.

"Fodder" refers particularly to foods or forages given to the animals (including plants cut and carried to them), rather than that which they forage for themselves. It includes hay, straw, silage, compressed and pelleted feeds, oils and mixed rations, and sprouted grains and legumes. Feed grains are the most important source of animal feed globally. The amount of grain used to produce the same unit of meat varies substantially. According to an estimate reported by the BBC in 2008, "Cows and sheep need 8 kg of grain for every 1 kg of meat they produce, pigs about 4 kg. The most efficient poultry units need a mere 1.6 kg of feed to produce 1 kg of chicken." Farmed fish can also be fed on grain, and use even less than poultry. The two most important feed grains are maize and soybean, and the United States is by far the largest exporter of both, averaging about half of the global maize trade and 40% of the global soya trade in the years leading up the 2012 drought. Other feed grains include wheat, oats, barley, and rice, among many others.

Traditional sources of animal feed include household food scraps and the byproducts of food processing industries such as milling and brewing. Material remaining from milling oil crops like peanuts, soy, and corn are important sources of fodder. Scraps fed to pigs are called slop, and those fed to chicken are called chicken scratch. Brewer's spent grain is a byproduct of beer making that is widely used as animal feed.

A pelleted ration designed for horses.

Compound feed is fodder that is blended from various raw materials and additives. These blends are formulated according to the specific requirements of the target animal. They are manufactured by feed compounders as *meal type*, *pellets* or *crumbles*. The main ingredients used in commercially prepared feed are the feed grains, which include corn, soybeans, sorghum, oats, and barley.

Compound feed may also include premixes, which may also be sold separately. Premixes are composed of microingredients such as vitamins, minerals, chemical preservatives, antibiotics, fermentation products, and other ingredients that are purchased from premix companies, usually in sacked form, for blending into commercial rations. Because of the availability of these products, a farmer who uses his own grain can formulate his own rations and be assured his animals are getting the recommended levels of minerals and vitamins, although he is still subject to the Veterinary Feed Directive.

According to the American Feed Industry Association, as much as $20 billion worth of feed ingredients are purchased each year. These products range from grain mixes to orange rinds to beet pulps. The feed industry is one of the most competitive businesses in the agricultural sector, and is by far the largest purchaser of U.S. corn, feed grains, and soybean meal. Tens of thousands of farmers with feed mills on their own farms are able to compete with huge conglomerates with national distribution. Feed crops generated $23.2 billion in cash receipts on U.S. farms in 2001. At the same time, farmers spent a total of $24.5 billion on feed that year.

In 2011, around 734.5 million tons of feed were produced annually around the world. The beginning of industrial-scale production of animal feeds can be traced back to the late 19th century, around the time advances in human and animal nutrition were able to identify the benefits of a balanced diet, and the importance of the role processing of certain raw materials played. Corn gluten feed was first manufactured in 1882, while leading world feed producer Purina Feeds was established in 1894 by William Hollington Danforth. Cargill, which was mainly dealing in grains from its beginnings in 1865, started to deal in feed at about 1884.

The feed industry expanded rapidly in the first quarter of the 20th century, with Purina expanding its operations into Canada, and opened its first feed mill in 1927 (which is still in operation). In 1928, the feed industry was revolutionized by the introduction of the first pelleted feeds - Purina Checkers.

Cattle eating a total mixed ration.

The US Animal Drug Availability Act 1996, passed during the Clinton era, was the first attempt in that country to regulate the use of medicated feed.

Forage

A herdsman from the Maasai people watches as his cattle
graze in the Ngorongoro crater, Tanzania.

"Forage" is plant material (mainly plant leaves and stems) eaten by grazing livestock. Historically, the term *forage* has meant only plants eaten by the animals directly as pasture, crop residue, or immature cereal crops, but it is also used more loosely to include similar plants cut for fodder and carried to the animals, especially as hay or silage.

Manufacture

Nutrition

In agriculture today, the nutritional needs of farm animals are well understood and may be satisfied through natural forage and fodder alone, or augmented by direct supplementation of nutrients in concentrated, controlled form. The nutritional quality of feed is influenced not only by the nutrient content, but also by many other factors such as feed presentation, hygiene, digestibility, and effect on intestinal health.

Feed additives provide a mechanism through which these nutrient deficiencies can be resolved effect the rate of growth of such animals and also their health and well-being. Even with all the benefits of higher quality feed, most of a farm animal's diet still consists of grain-based ingredients because of the higher costs of quality feed.

FEED MANAGEMENT

Feeding of Calves Colostrums

It will vary with the system followed, but whatever system may be practiced, the calf must receive the first milk which the cow gives after calving and is called colostrums. Be sure to feed the calf enough of colostrums between 2 to 2.5 liters daily for the first 3 days following its birth.

Any excess colostrums may be fed to other calves in the herd in amounts equal to the amount of whole milk normally fed. If possible where a cow is milked before calving, freeze some of the colostrums for later feeding to the calf. None of it should be wasted. The digestibility of colostrums increases when it is given at a temperature between 99 °F and 102 °F. The importance of colostrums can be felt more from the following virtues.

The protein of colostrums consists of a much higher proportion of globulin than doe's normal milk. The globulins are presumed to be the source of antibodies which aid in protecting the animal from many infections liable to affect it after birth. Gamma-globulin level in blood serum of neonatal calves is only 0.97 mg/ml at birth. It increase to 16.55 mg/ml level after first colostrums feeding at 12 hr and subsequently on the second day shows a peak of 28.18 mg/ml. This level more or less persists till the reti-culoendothelial system of the calf starts functioning to produce antibodies.

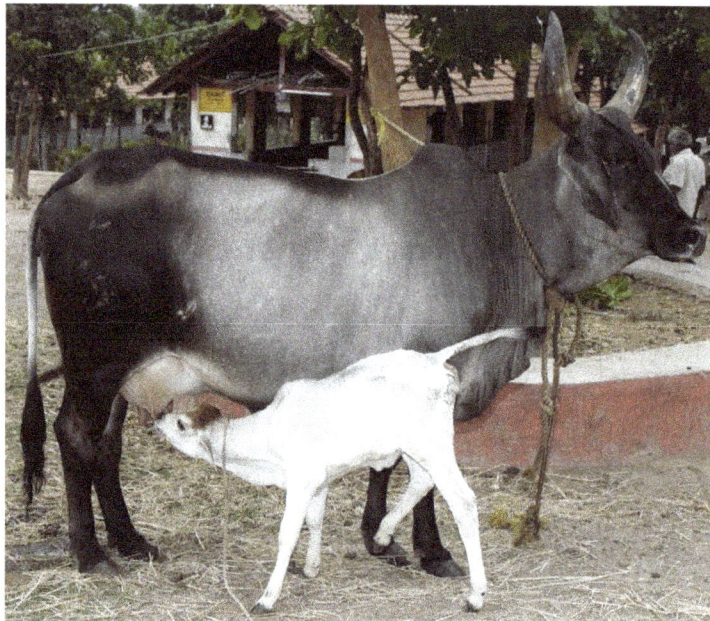

- The protein content of colostrums is 3 to 5 times as that of normal milk. It is also rich in some of the materials, of which copper, iron, magnesium and manganese are important.

- Colostrums contain 5-15 times the amount of Vitamin A- found in normal milk, depending upon the character of the ration given to the mother during the rest period.

- Colostrums is also superior to milk in having a considerably greater amount of several other vitamins which have been found essential in the growth of dairy calves, including riboflavin, choline, thiamine and pantothenic acid.

- Colostrums act as a laxative to free the digestive tract of faecal material.

Feeding Whole Milk

In feeding whole milk, calves may be fed as per feeding schedule. While feeding whole milk the following points should be remembered.

- As far as possible provide milk from the calf's mother.

- Feed milk immediately after it is drawn.

- The total amount of milk may be fed at 3 or 4 equal intervals up to the age of 7 days and then twice daily.

Feeding Skim Milk

On many farms, large quantities of separated milk are available for feeding to calves and other livestock. Excellent dairy calves can be raised by changing them from whole milk gradually after two weeks of their age. Here again the feeding schedule should be followed.

Feeding Dried Skim Milk, Whey or Buttermilk

The above dried products are mixed with water at the rate of 1 kg to 9 kg of water and then it is fed as skim milk. To avoid digestive troubles the mix should always be fed to calves after warming it up to 100 °F.

Feeding Calf Starters

Calf starter is a mixture consisting of ground farm grains, protein feeds and minerals, vitamins and antibiotics. After a calf attains the age of 2 weeks the amount of whole milk given to it may be cut down. One should then rub a small amount of starter on the calf's mouth, after each milk feeding for a few days when the calf will be accustomed to it. When they reach four months of age, one should then transfer the calves to a "growing" grain ration.

Feeding Grain Mixture

Feeding of Calves.

Better growth and greater resistance to calf ailments result from consumption of grain and milk by the calf then when the calf is fed only on milk. At the age of 7-15 days the feeding of grain mixtures may be started. In order to get calves accustomed to grain mixtures, place a small handful of grain mixture in the used pail. As the calf is finishing its milk it may consume a portion, or one may offer a little in the hand immediately after feeding milk.

Excessive protein rich grain mixture is not desirable as milk is already rich in proteins. A medium protein grain mixture is most suitable when milk is fed freely. A grain mixture of oats - 35 percent, linseed cake - 5 percent, bran - 30 percent, barley - 10 percent, groundnut cake - 20 percent may be fed to the calves. Another good mixture consists of ground maize - 2 parts, wheat bran - 2 parts.

Table: Feeding schedule for calves up to 6 months.

Age of calf	Approx. body weight (kg)	Quantity of milk (kg)	Quantity of calf starter (g)	Green grass (kg)
4 days to 4 weeks	25	2.5	Small qty.	Small qty.
4-6 weeks	30	3.0	50-100	Small qty.
6-8 weeks	35	2.5	100-250	Small qty.
8-10 weeks	40	2.0	250-350	Small qty.
10-12 weeks	45	1.5	350-500	1-0
12-16 weeks	55	-	500-750	1-2
16-20 weeks	65	-	750-1000	2-3
20-24 weeks	75	-	1000-1500	3-5

Calf starter is a highly nutritious concentrate mixture containing all the nutrients in proper proportion required for optimum growth and is used as a partial substitute for whole milk in the ration of calves. Since quality of protein is very important to calves until their rumen is fully functional, animal protein supplements such as fish meal should be included in calf starters. Urea should not be included in calf starters.

Feeding of Growing Animals from 6 Months Onwards

For calves below one year of age it is always desirable to give sufficient concentrates in addition to good roughage so that they make optimum growth. Feeding concentrate can be considerably reduced in the case of calves over one year of age fed on high quality roughage. A judicious mixture of roughage and concentrate is essential for obtaining optimum growth without undue fat deposition. From six months onwards, calves can be given the same type of concentrate mixture (14-16% Digestible Crude Protein and about 70% Total Digestible Nutrients) as used for adult cattle.

Table: Feeding schedule of growing animals from 6 months onwards.

Age (months)	Approximate body weight (kg)	Concentrate mixture (kg)	Grass (kg)
6-9	70-100	1.5-1.75	5-10
9-15	100-150	1.75-2.25	10-15
15-20	150-200	2.25-2.50	15-20
Above 20	200-300	2.50-2.75	15-20

Table: Feeding schedules for dairy animals (quantity in kg).

			Cross Breed Cow		
S. No.	Type of animal	Feeding during	Green Fodder	Dry Fodder	Concentrate
1.	6 to 7 liters milk per day	Lactation days	20 to 25	5 to 6	3.0 to 3.5
		Dry days	15 to 20	6 to 7	0.5 to 1.0
2.	8 to 10 liters milk per day	Lactation days	25 to 30	4 to 5	4.0 to 4.5
		Dry days	20 to 25	6 to 7	0.5 to 1.0

Feeding of Lactating Cow

Proper feeding of dairy cattle should envisage minimum wastage of nutrients and maximum returns in respect of milk produced. A concentrate mixture made up of protein supplements such as oil cakes, energy sources such as cereal grains (maize, jowar), tapioca chips and laxative feeds such as brans (rice bran, wheat bran, gram husk) is generally used. Mineral mixture containing major and all the trace elements should be included at a level of 2 percent.

Table: Feeding schedule for different classes of adult cows (approximate body weight-250 kg).

Category	When green grass is plenty		When paddy straw is the major roughage		
	Concentrate mixture (Kg)	Green Grass (kg)	Concentrate Mixture (kg)	Green Grass (kg)	Paddy Straw (kg)
Dry cows	-	25 – 30	1.25	5.0	5 – 6
Milking	1 kg for every 2.5 - 3.0 kg of milk	30	1.25 + 1 kg for every 2.5 - 3.0 kg of milk	5.0	5 – 6
Pregnant	Production Allowance + 1 to 1.5 kg from 6th month of pregnancy	25 - 30	Maintenance + production + 1 to 1.5 kg from 6th month of pregnancy	5.0	5 - 6

The total dry matter requirement of cattle is around 2-3 % of their body weight though high yielding animals may eat at a rate more than 3%. Such factors as climate, processing of feeds, palatability etc. influence the dry matter consumption. Good quality grasses (Guinea, Napier etc.) with a minimum of 6 % crude protein on dry matter basis alone can form maintenance ration of a cow of average size. But it is possible to maintain milk production of up to 3-4 kg with grass-legume fodder.

Feeding of Bulls

Male calves to be reared as future breeding bulls, should be fed on a higher plane of nutrition than female calves.

Table: Feeding schedule of bull.

Body weight (kg)	Concentrate mixture (kg)	Green grass (kg)
400-500	2.5-3	20-25

A bull in service should be given good quality roughage with sufficient concentrates. Too much roughage feeding should be avoided as it makes the bull paunchy and slow in service. A large concentrate allowance may make the bull too much fatty and less virile.

Maternity Pen or Calving Boxes

In the large farms calving boxes are provided for cows nearing parturition. The cows are transferred into these pens 2-3 weeks before expected date of calving. Each calving pen should be about 3 x 4 m for covered area and another 4 x 5 m for open paddocks. A manger and water trough of proper size should be constructed in each pen. The covered area shall have 1.25 m high walls all round. A 1.2 m wide gate opening into the open lot is also provided. The floor shall be of cement or brick on edge with slope towards drain. In large farms the member of calving boxes required is about 5% of number of breed able stock in the farm. These pens are located nearer to quarters of the farmer/milking barn to monitor pregnant rows. Adequate lighting should be made.

Nutrient Requirement

- Concentrate must be feed individually according to production requirements.

- Good quality roughage saves concentrates. Approximately 20 kg of grasses (guinea, napier, etc.) or 6-8 kg legume fodder (cowpea, lucerne) can replace 1 kg of concentrate mixture (0.14-0.16 kg of DCP) in terms of protein content.

- 1 kg straw can replace 4-5 kg of grass on dry matter basis. In this case the deficiency of protein and other nutrients should be compensated by a suitable concentrate mixture.

- Regularity in feeding should be followed. Concentrate mixture can be fed at or preferably before milking – half in the morning and the other half in the evening – before the two milkings. Half the roughage ration can be fed in the forenoon after watering and cleaning the animals. The other half is fed in the evening, after milking and watering. High yielding animals may be fed three times a day (both roughage and concentrate). Increasing the frequency of concentrate feeding will help maintain normal rumen motility and optimum milk fat levels.

- Over-feeding concentrates may result in off feed and indigestion.

- Abrupt change in the feed should be avoided.

- Grains should be ground to medium degree of fineness before being fed to cattle.

- Long and thick-stemmed fodders such as Napier may be chopped and fed.

- Highly moist and tender grasses may be wilted or mixed with straw before feeding. Legume fodders may be mixed with straw or other grasses to prevent the occurrence of bloat and indigestion.

- Silage and other feeds, which may impart flavour to milk, may be fed after milking. Concentrate mixture in the form of mash may be moistened with water and fed immediately. Pellets can be fed as such.

- All feeds must be stored properly in well-ventilated and dry places. Mouldy or otherwise damaged feed should not be fed.

- For high yielding animals, the optimum concentrate roughage ratio on dry matter basis should be 60:40.

Table: Nutrients required for maintenance of adult cattle per head per day (growth rate- 550g per day).

Live weight (kg)	Dry Matter (kg)	Digestible Crude Protein (g)	Total Digestible Nutrients (kg)	Calcium (g)	Phosphorus (g)
250	4-5	140	2.2	25	17
300	5-6	168	2.65	25	17
350	6-7	195	3.10	25	17
400	7-8	223	3.55	28	20
450	8-9	250	4.00	31	23
500	9-10	278	4.45	31	23
550	10-11	310	4.90	31	23
600	11-12	336	5.35	31	23

Straw can form the roughage in the absence of grasses and in such cases concentrates should be given for maintenance. For lactating cows, 1kg of concentrate mixture (compounded feed) (0.14-0.16 kg DCP and 0.70 kg TDN) may be required for every 2.5 – 3.0 kg of milk over and above the maintenance allowance. After parturition, the cow should be given the same type of feed and the same quantity as before and the concentrate allowance should be only gradually increased to avoid digestive troubles like acidosis, indigestion, etc.

In the case of young cross-bred cows below four years of age to meet the needs for growth, it is desirable to give additional concentrate allowance at the rate of 1kg for animals in first lactation and 0.5 kg in the second lactation over and above the maintenance and production needs. Milking animals should always have free access to clean fresh drinking water.

Table: Bureau of Indian standards specification for mineral mixture for cattle.

S. No	Characteristics	Type I (with salt)	Type II (without salt)
1.	Moisture, percent by mass, Max.	5	5
2.	Calcium, percent by mass, Min.	18	23

3.	Phosphorus, percent by mass, Min.	9	12
4.	Magnesium, percent by mass, Min.	5	6.5
5.	Salt (Chloride as Sodium Chloride), percent by mass, Min.	22	-
6.	Iron, percent by mass, Min.	0.4	0.5
7.	Iodine (as KI), percent by mass.	0.02	0.026
8.	Copper, percent by mass, Min.	0.06	0.077
9.	Manganese, percent by mass, Min.	0.10	0.12
10.	Cobalt, percent by mass, Min.	0.009	0.012
11.	Fluorine, percent by mass, Max.	0.05	0.07
12.	Zinc, per cent by mass, Min.	0.30	0.38
13.	Sulphur, percent by mass, Max.	0.40	0.50
14.	Acid insoluble ash, percent by mass	3.00	2.50

LIVESTOCK MANAGEMENT

Management of livestock must take into account variable seasonal factors, fluctuating markets and declining terms of trade. The most successful producers have a good knowledge of market requirements, matching product quality to suit. There are many factors that can determine the productivity and profitability of a livestock enterprise. These include the supply and quality of feedstuffs, the use of the most appropriate genetics, ensuring high health standards, optimising housing or environmental conditions, meeting quality assurance requirements, and having a sound knowledge of market requirements. This requires good communication along the value chain.

Livestock Management Tips

The development and application of modern technology have enhanced the health of livestock or farm animals considerably. Some of the benefits include - high profits and well-organized livestock management. Moreover, the growing use of precise, robust and consistent livestock production tools has improved the monitoring, administration and movement of cattle, which has resulted in momentous productivity gains across the farming business.

Here are few best practices to make your livestock management easy and efficient so that your cattle give better results:

Provide Nutritional Diet

Always select feedstuffs that are loaded with nutrients so that your cattle reach their best levels of production. A smaller amount of red meat and more feed variability will maximize the production of both - milk and meat. Feeding animals with less human food and letting them graze pastures that have rich-fibre content will improve their overall nutrition ingestion and resistance to unbearable diseases.

Give Right Food Supplements

Supplements have proved to improve animal health as well as productivity by encouraging the growth of helpful microbes in the rumen. Red clover is famous for an enzyme called Trifolium pratense that increases the capability of your livestock to take in dietary protein. The presence of clover in the feed helps in improving milk production and also increases the appetite of livestock. Tar brush supplements avoid gastrointestinal acidosis in livestock and reduce the release of green-houses gasses. Azolla Caroliniana (a water fern grown in ponds) offers added protein to animals that are lacking in protein. By including supplements to your farm animals daily feed is the best way to increase productivity of ruminant animals.

Use Technology

Technology facilitates reliable and accurate examining of animal health by providing appropriate and accurate interpretations in the form of structural figures. The arrival of imaging tools and live-stock scales permit remote monitoring of animal health and help farmers in making informed decisions regarding rearing and feeding patterns. You will get several makes and models of livestock scales on the market depending on your resources. These automated cattle scales are long-lasting, cost-effective and accurate for:

- Checking animal health,

- Knowing their exact weight before breeding,

- Evaluating the conversion of feed,

- Assessing their performance.

Track Animal Performance

If you track the performance of livestock, you can easily recognize healthy breeds that do better and pick the unproductive breeds and use them for other things. Checking animal weight facilitates early recognition of diseases and prevents dangerous disease from spreading to other animals in the farm. Monitoring livestock will also help in selecting right weaning time and choosing animals for cross breeding.

Take Suitable Precautions

Animals that are imported from moderate climates often lack resistance to heat, humidity, ticks, parasites and tropical diseases hence it is essential to keep them in dirt free stalls where they stay away from disease vectors. Instead of letting your farm animals to graze, you must cut the silage and give them at the stalls. You can also purchase imported feed that assures quality.

Customize Practices Favorable to Local Climate

A number of people depend on livestock for livelihood but the benefits of rearing cattle could stop if the customary farming practices and conventional grazing are restored with industrial systems that do not take natural factors into deliberation. Altering farm practices for profits will work only when the local breeds, resources as well as feedstuffs are utilized and wastage is reduced. Local breeds when given a proper diet along with fresh water and supplements will stay healthy and be productive and profitable for the farmers.

LIVESTOCK DISEASE MANAGEMENT

Livestock systems in developing countries are characterised by rapid change, driven by factors such as population growth, increases in the demand for livestock products as incomes rise, and urbanisation. Climate change is adding to the considerable development challenges posed by these drivers of change. The increasing frequency of heat stress, drought and flooding events could translate into the increased spread of existing vector-borne diseases and macro-parasites, along with the emergence of new diseases and transmission models. Appropriate sustainable livestock management practices are required so that livestock keepers can take advantage of the increasing demand for livestock products (where this is feasible) and protect their livestock assets in the face of changing and increasingly variable climates.

Livestock diseases contribute to an important set of problems within livestock production systems. These include animal welfare, productivity losses, uncertain food security, loss of income and negative impacts on human health. Livestock disease management can reduce disease through improved animal husbandry practices. These include: controlled breeding, controlling entry to farm lots, and quarantining sick animals and through developing and improving antibiotics, vaccines and diagnostic tools, evaluation of ethno-therapeutic options, and vector control techniques.

Livestock disease management is made up of two key components:

- Prevention (biosecurity) measures in susceptible herds.

- Control measures taken once infection occurs.

The probability of infection from a given disease depends on existing farm practices (prevention) as well as the prevalence rate in host populations in the relevant area. As the prevalence in the area increases, the probability of infection increases.

Prevention Measures

Preventing diseases entering and spreading in livestock populations is the most efficient and cost-effective way of managing disease. While many approaches to management are disease specific, improved regulation of movements of livestock can provide broader protection. A standard disease prevention programme that can apply in all contexts does not exist. But there are some basic principles that should always be observed. The following practices aid in disease prevention:

- Elaboration of an animal health programme.

- Select a well-known, reliable source from which to purchase animals, one that can supply healthy stock, inherently vigorous and developed for a specific purpose. New animals should be monitored for disease before being introduced into the main flock.

- Good hygiene including clean water and feed supplies.

- Precise vaccination schedule for each herd or flock.

- Observe animals frequently for signs of disease, and if a disease problem develops, obtain an early, reliable diagnosis and apply the best treatment, control, and eradication measures for that specific disease.

- Dispose of all dead animals by burning, deep burying, or disposal pit.

- Maintain good records relative to flock or herd health. These should include vaccination history, disease problems and medication.

Surveillance and Control Measures

Disease surveillance allows the identification of new infections and changes to existing ones. This involves disease reporting and specimen submission by livestock owners, village veterinary staff, district and provincial veterinary officers. The method used to combat a disease outbreak depends on the severity of the outbreak. In the event of a disease outbreak the precise location of all livestock is essential for effective measures to control and eradicate contagious viruses. Restrictions on animal movements may be required as well as quarantine and, in extreme cases, slaughter. Figures are photos illustrating the holistic approaches to livestock disease prevention and control.

Holistic approaches to disease prevention control (woman and man participants in rural training course in learning how to improve health of their goats - Sudan).

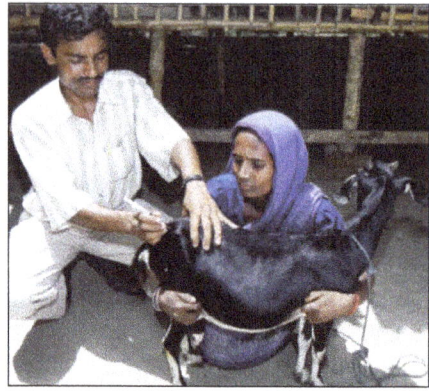
Holistic approaches to disease prevention control (Man immunising goat held by woman - Bangladesh).

The major impacts of climate change on livestock diseases have been on diseases that are vector-borne. Increasing temperatures have supported the expansion of vector populations into cooler areas. Such cooler areas can be either higher altitude systems (for example, livestock tick-borne diseases) or more temperate zones (for example, the outbreak of bluetongue disease in northern Europe). Changes in rainfall pattern can also influence an expansion of vectors during wetter years and can lead to large outbreaks. Climate changes could also influence disease distribution indirectly through changes in the distribution of livestock. Improving livestock disease control is therefore an effective technology for climate change adaptation.

Advantages of the Technology

Benefits of livestock disease prevention and control include: higher production (as morbidity is lowered and mortality or early culling is reduced), and avoided future control costs. When farmers mitigate disease through prevention or control, they benefit not just themselves but any others at risk of adverse outcomes from the presence of disease on that operation. At-risk populations include residents, visitors and consumers. The beneficiaries might also include at-risk wildlife populations surrounding the farm that may have direct or indirect contact with livestock or livestock-related material.

Disadvantages of the Technology

Management options may interact, so the use of one option may diminish the effectiveness of another. Another critical issue is the long-term sustainability of currently used strategies. Chemical intervention strategies such as antibiotics or vaccines are not biologically sustainable. Animals develop resistance to drugs used to control certain viruses and with each new generation of vaccine a new and more virulent strain of the virus can arise. Small-scale producers may be negatively affected by livestock disease management if the full cost of the disease management programme is directly passed onto them with no subsidy from the government.

Financial Requirements and Costs

Livestock disease management costs include: testing and screening, veterinary services, vaccines, training of livestock keepers and veterinary staff, and perhaps changes to practices and facilities to reflect movement restrictions and quarantines when animals are added to the herd.

Control of Mastitis

A low-cost technology applicable to a wide range of livestock (cattle, sheep and goats) is the control of mastitis. Mastitis is an infectious disease caused by pathogenic micro-organisms due to inadequate milking practices or blows to the udders. It is one of the diseases that cause the most financial losses in milk production. In conditions of increasing climate variability, emergence of new pests and diseases can introduce invasive organisms to the livestock environment. It is therefore essential that livestock farmers are able to identify and prevent mastitis in order to maintain healthy animals that, in turn, are more capable of withstanding adverse weather conditions such as prolonged droughts or severe frosts.

Information and monitoring requirements for the control of mastitis include:

- Producer training on testing and diagnosing mastitis, hygienic milking practices, teat sealing, treatment of clinical mastitis, control records.

- Organisations or institutions must have extension farmers or technicians who are trained in the mastitis control process.

- Monitoring and regular check-ups are necessary for the prevention of mastitis.

The following is also required in the application of this technology:

- The California Mastitis Test (CMT) or black background rate is very easy for farmers to use as readings are immediate and low cost.

- Teat sealant to protect the udder against mastitis germs.

- Clean and disinfected containers, cloths and mechanical milking machines.

- Milking records which should contain basic information like the name of the animal, the date, the name of the person milking the animal, the rooms examined, evidence of mastitis, density and acidity of the milk.

Institutional and organisational requirements must also be taken into account: health care institutions and producers' organisations should carry out sanitation campaigns, hold training workshops and provide technical assistance on the control of mastitis, using adequate informative materials like easy-to-read leaflets and flyers that the cattle farmers can understand and follow. Costs and financial requirements are relatively low. The CMT costs about US $25 and can last about six months for an average of three cows per farmer. The teat sealant costs about US $30.

In a project implemented by Practical Action Latin America in San Miguel province in the Cajamarca region of Peru, two Livestock Services Centres were formed. These centres comprised extension farmers who had participated in a training programme on livestock management, animal health, animal feed, genetic improvement, business management, and information and communication technologies. This enabled them to provide training and technical assistance in their 22 settlements or communities. At the present time, 22 extension farmers are providing more than 450 services, dealing with problems affecting the dairy cattle and providing training in their communities on mastitis control and milk analysis; hygienic milking; milk control records and dairy cattle management. This mastitis control practice was applied in 50 per cent of the dairy farms, improving the quality of the milk and increasing production by 10 per cent.

Prevention and control costs are generally evaluated against expected financial losses resulting from a disease outbreak in a cost-benefit analysis. The assumption is that increased prevention and control costs lower the expected losses by diminishing the expected scale of an infection. McInerney et al present the problem graphically as a cost minimisation problem:

$$\min C = L + E$$

where C is total annual disease cost, L is the value of output losses, and E is the control expenditures (which themselves are a function of inputs purchased for control).

Institutional and Organisational Requirements

Countries should cooperate in programmes against trans-boundary disease either through formally formed organisations or networks. Neighbouring countries often have similar production systems and disease risk profiles and will be more likely to be affected by similar climate change impacts in livestock disease. There will be mutual benefits and cost savings through joint preparedness planning. Public policies range from bounties/indemnities for infected livestock to required herd depopulation and farm decontamination, to decentralisation programmes for provision of veterinary services and drug supplies. Livestock and animal health policy should be oriented to both the commercial and pastoral sectors and include pro-poor interventions to support the most vulnerable populations. Government investments in infrastructure (including early warning systems, roads, abattoirs, holding pens, processing plants, air freight/ports and so on), systematic vaccination, and in research and development can all contribute to providing an enabling environment for effective livestock disease management. Removing or introducing subsidies for improved management, insurance systems and supporting income diversification practices could benefit adaptation efforts.

In order for producers to make decisions regarding disease management, they must understand the options that they have. These options depend on disease biology, prevention techniques, tests for infection and their costs, treatments available, market reactions, as well as industry and government programmes and policies. Disease biology includes transmission modes and rates, disease evolution (for example, length of time to infectious period), production losses associated with the disease, and mortality rate (where applicable).

Practical training for farmers should include:

- Principles of anatomy and physiology of the livestock animals.

- Principles of nutrition and pasture ecology.

- Animal diseases of local importance: clinical and post mortem signs, epidemiology, prevention, treatment. Applying first aid, the use of basic veterinary medicines (wound treatments, dips, anthelmintics, antibiotics, trypanocides, babesiacides, vaccines, care and storage of medicines and vaccines, and the use and care of syringes).

- The basic principles of sero-surveillance campaigns — how to draw blood and store sera.

Modelling disease outbreaks and spread can provide valuable information for the development of management strategies. Modelling involves studying disease distribution and patterns of spread to determine the scale of a problem. This information is used to develop a model that can predict the spread of disease. Disease modelling requires prior knowledge of animal population distributions and ecology, diseases present and methods of disease transmission. Modelling can be used to assess potential disease impacts and develop contingency plans.

Geographic Information System (GIS) software can play a key role in livestock disease management. The main advantage of GIS software is not just that the user can see how a disease is distributed geographically, but also that an animal disease can be viewed against other information. For example, maps that show possible impacts of climate change on rainfall patterns, crop yields and flooding. The disease presence can then be related to these factors and more easily appreciated visually. This is important in relation to managing and responding to the changes in distribution of diseases due to changing climate.

Role of Indigenous Knowledge in Livestock Disease Management under Climate Change

Indigenous knowledge about livestock disease management has been shown, in certain cases, to be cost-effective, sustainable, environmentally friendly and practical. Practices include:

- Utilisation of local plant remedies for prevention and cure of diseases.

- Avoiding certain pastures at particular times of the year; and not staying too long in one place to avoid parasite build-up.

- Lighting smoke fires to repel insects, especially tsetse flies.

- Mixing species in the herd to avoid the spread of disease.

- Avoiding infected areas or moving upwind of them; spreading livestock among different herds to minimise risks; and quarantining sick animals.

- Selective breeding: As an example from the arid south of Zambia, restocking and promoting the rearing of drought-tolerant goat breeds are adaptive measures already being undertaken.

Barriers to Implementation

A lack of strong institutions and political will to monitor disease status effectively can produce a considerable barrier to livestock disease management. Difficulties in eradication of disease may also be exacerbated by many small-scale and backyard producers, infected wildlife, smuggling, and cockfighting. If there is no compensation for stamping out disease through slaughter, then producers, particularly small-scale producers, may be reluctant to participate. If they do participate it may mean that they no longer can afford to produce.

Opportunities for Implementation

Where the disease organism has built up resistance against vaccines or the animal has built resistance against the disease there is an opportunity for incorporating simple, high-tech genetic approaches such as selective breeding. National planning for livestock disease management also presents an opportunity to improve agricultural support services in rural areas and to incorporate indigenous knowledge into formal prevention and control plans, thereby unlocking the potential of low-cost interventions and disseminating information on traditional lessons and experiences to a wider audience. Trans-border collaboration can provide an opportunity to strengthen veterinary services and can improve the effectiveness of disease management programmes through harmonisation of prevention and control measures, such as disease reporting and surveillance.

References

- Livestock-farming, topic: britannica.com, Retrieved 14 June, 2019

- "Health and Consumer Protection - Scientific Committee on Animal Health and Animal Welfare - Previous outcome of discussions (Scientific Veterinary Committee) - 17". Archived from the original on May 22, 2013. Retrieved September 6, 2015

- Livestock-management, livestock-animals: agric.wa.gov.au, Retrieved 15 July, 2019

- Foer, Jonathan Safran (2010). Eating Animals. Hachette Book Group USA. ISBN 978-0316127165. OCLC 669754727

- Important-livestock-management-tips-for-farmers, animal-husbandry: krishijagran.com, Retrieved 16 August, 2019

- Brian Machovia, K. J. Feeley, W. J. Ripple, "Biodiversity conservation: The key is reducing meat consumption", Science of the Total Environment, 536, December 1, 2015, 419–431. Doi:10.1016/j.scitotenv.2015.07.022 PMID 26231772

- Livestock-disease-management: climatetechwiki.org, Retrieved 17 January, 2019

- Chopin T, Buschmann AH, Halling C, Troell M, Kautsky N, Neori A, Kraemer GP, Zertuche-Gonzalez JA, Yarish C and Neefus C. 2001. Integrating seaweeds into marine aquaculture systems: a key toward sustainability. Journal of Phycology 37: 975–986

Poultry Farming and Management

The form of animal husbandry which deals with raising domesticated birds for eggs or meat is known as poultry farming. Poultry management deals with the production practices for maximizing the efficiency of production as well as handling poultry diseases. The topics elaborated in this chapter will help in gaining a better perspective about the different practices associated with poultry farming and management.

Poultry farming is the raising of birds domestically or commercially, primarily for meat and eggs but also for feathers. Chickens, turkeys, ducks, and geese are of primary importance, while guinea fowl and squabs (young pigeons) are chiefly of local interest.

Commercial Production

Feeding

Commercial poultry feeding is a highly perfected science that ensures a maximum intake of energy for growth and fat production. High-quality and well-balanced protein sources produce a maximum amount of muscle, organ, skin, and feather growth. The essential minerals produce bones and eggs, with about 3 to 4 percent of the live bird being composed of minerals and 10 percent of the egg. Calcium, phosphorus, sodium, chlorine, potassium, sulfur, manganese, iron, copper, cobalt, magnesium, and zinc are all required. Vitamins A, C, D, E, and K and all of the B vitamins are also required. Antibiotics are widely used to stimulate appetite, control harmful bacteria, and prevent disease. For chickens, modern rations produce about 0.5 kg (1 pound) of broiler on about 0.9 kg (2 pounds) of feed and a dozen eggs from 2 kg (4.5 pounds) of feed.

Management

A carefully controlled environment that avoids crowding, chilling, overheating, or frightening is almost universal in poultry farming. Cannibalism, which expresses itself as toe picking, feather picking, and tail picking, is controlled by debeaking at one day of age and by other management practices. The feeding, watering, egg gathering, and cleaning operations are highly mechanized. Birds are usually housed in wire cages with two or three animals per cage, depending on the species and breed, and three or four tiers of cages superposed to save space. Cages for egg-laying birds have been found to increase production, lower mortality, reduce cannibalism, lower feeding requirements, reduce diseases and parasites, improve culling, and reduce both space and labour requirements.

Poultry breeding is an outstanding example of the application of basic genetic principles of inbreeding and crossbreeding as well as of intensive mass selection to effect faster and cheaper gains in meat and maximum egg production for the egg-laying strains. Maximum use of heterosis, or hybrid vigour, through incrosses and crossbreeding has been made. Rapid and efficient weight gains and high-quality, plump, meaty carcasses have been achieved thereby.

Single-comb White Leghorn hens housed for egg production in a multitiered layer house.

Among the world's agricultural industries, chicken breeding in the U.S. is one of the most advanced. Intensive nutritional research and application, highly improved breeding stock, intelligent management, and scientific disease control have gone into the effort to give a modern broiler (meat chicken) of uniformly high quality produced at ever-lower cost. A modern broiler chick can reach a 2.3 kg (5-pound) market weight in five weeks, compared with the four months that were required in the mid-20th century. Additionally, annual egg production per hen has increased from about 100 in 1910 to over 300 in the early 21st century.

Diseases

Poultry are quite susceptible to a number of diseases; some of the more common are fowl typhoid, pullorum, fowl cholera, chronic respiratory disease, infectious sinusitis, infectious coryza, avian infectious hepatitis, infectious synovitis, bluecomb, Newcastle disease, fowl pox, avian leukosis complex, coccidiosis, blackhead, infectious laryngotracheitis, infectious bronchitis, and erysipelas. Strict sanitary precautions, the intelligent use of antibiotics and vaccines, and the widespread use of cages for layers and confinement rearing for broilers have made it possible to effect satisfactory disease control.

Parasitic diseases of poultry, including hexamitiasis of turkeys, are caused by roundworms, tapeworms, lice, and mites. Again, modern methods of sanitation, prevention, and treatment provide excellent control.

Types of Poultry

Chickens

Mass production of chicken meat and eggs began in the early 20th century, but by the middle of that century meat production had outstripped egg production as a specialized industry. The market for chicken meat has grown dramatically since then, with worldwide exports reaching nearly 12.5 million metric tons (about 13.8 million tons) by the early 21st century.

The breeds of chickens are generally classified as American, Mediterranean, English, and Asiatic. While there are hundreds of breeds in existence, commercial facilities rely on only a select few that meet the rigorous demands of industrial production. The single-comb White Leghorn, a Mediterranean breed widely used throughout the global egg industry, is a prolific layer that quickly reaches sexual maturity. The Cornish Cross, a hybrid of Cornish and White Rock, is one of the most-common breeds for industrial meat production and is esteemed for its compact size and rapid, efficient growth.

Egg-producing hens (Gallus gallus) in an industrial henhouse. Under ideal lighting and heating conditions, female chickens may produce one egg every 23–26 hours.

Small farms and backyard flocks utilize a much wider variety of breeds and hybrids. Common American breeds include the Plymouth Rock, the Wyandotte, the Rhode Island Red, and the New Hampshire, all of which are dual-purpose breeds that are good for both eggs and meat. The Asiatic Brahma, thought to have originated in the United States from birds imported from China, is popular for both its meat and its large brown eggs.

Rhode Island Red rooster. Over the 7,400-year history of chicken (Gallus gallus) domestication, roosters (male chickens) have been used as fighting animals as well as for breeding and meat production.

Turkeys

After World War II, turkey production became highly specialized, with larger flocks predominating. Turkeys are raised in great numbers in Canada where their ancestors still live wild, as well as in some parts of Europe, the United States, Mexico, and Brazil. A hybrid white turkey dominates commercial production, while the Broad Breasted Bronze, the Broad Breasted White, the White Holland, and the Beltsville Small White are common breeds for smaller farms. In breeding flocks, one tom is required per 8 or 10 hens, though the modern hybrid turkey is too large for natural breeding and must be artificially inseminated.

Turkey, Midget White turkeys, a domestic breed of small stature.

Modern turkey breeding and farming practices have significantly reduced both the amount of feed and the time required to produce a pound of turkey meat. In 12–14 weeks a hen turkey eats about 16 kg (35 pounds) of feed and reaches 6–9 kg (14–20 pounds). Toms require some 36 kg (80 pounds) of feed to reach a market weight of 16–19 kg (35–42 pounds) in 16–19 weeks. Smaller turkey broilers are marketed from 12 to 15 weeks of age. Turkeys can be raised on open land with automatic waterers, self-feeders, range shelters, heavy fencing, and rotated pastures; however, they are often "grown out" commercially in rearing houses under environmentally controlled conditions.

Ducks and Geese

Duck raising is practiced on a limited scale in nearly all countries, usually as a small-farm enterprise, though some commercial plants do exist. Ducks are easily transported, can be raised in close confinement, and convert some waste products and scattered grain (e.g., by gleaning rice fields) to nutritious and very desirable eggs and meat. Khaki Campbell and Indian Runner ducks are prolific layers, each averaging 300 eggs per year. The Pekin duck, one of the most popular breeds in the United States, is used for both egg and meat production. Although the white-fleshed Aylesbury was once the favoured meat duck in England, disease and market competition from the yellow-fleshed Pekin duck have led to its decline.

Domesticated Pekin ducks (Anas platyrhynchos domestica). Pekin ducks are prolific
egg layers and are one of the most popular breeds in the United States.

Goose raising is often a minor farm enterprise, though some European countries have large-scale goose-production facilities. The two outstanding meat breeds are the Toulouse, predominantly gray in colour, and the Embden (or Emden), which is white. The birds are raised for meat and eggs as well as for their down feathers. Geese do not appear to have attracted the attention of geneticists on the same scale as the meat chicken and the turkey, and no change in the goose industry comparable to that in the others has occurred or seems to be in prospect. In some commercial plants, geese are fattened by a special process of force-feeding, resulting in a considerable enlargement of their livers, which are sold as the delicacy foie gras.

Flock of domestic geese (Anser anser domesticus).
Domestic geese are often kept as poultry for their eggs, meat, and feathers.

Guinea Fowl and Squabs

Guinea fowl are raised as a sideline on a few farms in many countries and are eaten as gourmet items. In Italy there is a fairly extensive industry. The birds are often raised in yards with open-fronted shelters, and a number of varieties and species are utilized throughout the world. Guinea fowl are marketed in England at 16–18 weeks of age and in the United States at about 10–12 weeks. The market weight is usually about 1–1.5 kg (2.5–3.5 pounds), but food conversion is poor.

Guinea fowl. Helmeted guinea fowl (Numida meleagris).

Pigeons are raised not only as messengers and for sport but also for the meat of their squabs (nestlings). Squab production, carried on locally, is rare in most countries with established poultry industries, though the meat is often marketed as a gourmet item.

POULTRY MANAGEMENT

Poultry management usually refers to the husbandry practices or production techniques that help to maximize the efficiency of production. Sound management practices are very essential to optimize production. Scientific poultry management aims at maximizing returns with minimum investment.

Brooder Management

Brooder house should be draft-free, rain-proof and protected against predators. Brooding pens should have windows with wire mesh for adequate ventilation. Too dusty environment irritates the respiratory tract of the chicks. Besides dust is one of the vehicles of transmission of diseases. Too much moisture causes ammonia fumes which irritate the respiratory tract and eyes. Good ventilation provides a comfortable environment without draft.

Sanitation and Hygiene

All movable equipments like feeders, waterers and hovers should be removed from the house, cleaned and disinfected. All litters are to be scraped and removed. The interior as well as exterior of the house should be cleaned under pressure. The house should be disinfected with any commercial disinfectant solution at the recommended concentration. Insecticide should be sprayed to avoid insect threat. Malathion spray/blow lamping or both can be used to control ticks and mites. New litter should be spread after each cleaning. The insecticides if necessary should be mixed with litter at recommended doses.

Litter

Suitable litter material like saw dust and paddy husk should be spread to a length of 5 cm depending upon their availability and cost. Mouldy material should not be used. The litter should be stirred at frequent intervals to prevent caking. Wet litters if any should be removed immediately and replaced by dry new litter. This prevents ammoniacal odour.

Brooding Temperature

Brooder

Heating is very much essential to provide right temperature in the brooder house. Too high or too low a temperature slows down growth and causes mortality. During the first week the temperature should be 95 °F (35 °C) which may be reduced by 5 °F per week during each successive week till 70 °F (21·1 °C). The brooder should be switched on for at least 24 hours before the chicks arrive. As a rule of thumb the temperature inside the brooder house should be approximately 20 °F (-6·7 °C) below the brooder temperature. Hanging of a maximum and minimum thermometer in each house is recommended to have a guide to control over the differences in the house temperature.

The behavior of chicks provides better indication of whether they are getting the desired amount of heat. When the temperature is less than required, the chicks try to get closer to the source of heat and huddle down under the brooder. When the temperature is too high, the chicks will get away from the source of heat and may even pant or gasp. When temperature is right, the chicks will be found evenly scattered. In hot weather, brooders are not necessary after the chicks are about 3 weeks old. Several devices can be used for providing artificial heat. Hover type electric brooders are by far the most common and practical these days. The temperature in these brooders is thermostatically controlled. Many a times the heat in the brooder house is provided by use of electric bulbs of different intensities. Regulation of temperature in such cases is difficult although not impossible. Infrared lamps are also very good for brooding. The height and number of infra-red lamps can be adjusted as per temperature requirement in the brooder house.

Brooder Space

Brooder space of 7 to 10 sq inch (45-65 cm²) is recommended per chick. Thus a 1·80 m hover can hold 500 chicks. When small pens are used for brooding, dimension of the house must be taken into consideration as overcrowding results in starve-outs, culls and increase in disease problems.

Brooder Guard

To prevent the straying of baby chicks from the source of heat, hover guards are placed 1·05 to 1·50 m from the edge of hover. Hover guard is not necessary after 1 week.

Floor Space

Floor space of 0·05 m² should be provided per chick to start with, which should be increased by 0·05 m2 after every 4 weeks until the pullets are about 20 weeks of age. For broilers at least 0·1 m² of floor space for female chicks and 0·15 m² for male chicks should be provided till 8 weeks of age. Raising broiler pullets and cockerel chicks in the separate pens may be beneficial.

Water Space

Plentiful of clean and fresh water is very much essential. A provision of 50 linear cm of water space per 100 chicks for first two weeks has to be increased to 152-190 linear cm at 6 to 8 weeks. When changing from chick fountain to water trough the fountains are to be left in for several days till the chicks have located the new water source. Height of the waterers should be maintained at 2·5 cm above the back height of the chicks to reduce spoilage. Antibiotics or other stress medications may be added to water if desired. All waterers should be cleaned daily. It may be desirable to hold a few chicks one at a time and teach them to drink.

PRINCIPLES OF POULTRY HUSBANDRY

There are a number of requirements by which animals should be managed so that the best performance is achieved in a way acceptable to those responsible for the care of the animals and to the community generally. These requirements are the keys to good management and may be used to test the management of a poultry enterprise in relation to the standard of its management. These requirements are also called Principles.

The importance of each Principle changes with the situation and thus the emphasis placed on each may alter from place to place and from time to time. This means that, while the Principles do not change, the degree of emphasis and method of application may change. Every facet of the poultry operation should be tested against the relevant principle(s).

Quality and Class of Stock

If the enterprise is to be successful it is necessary to use stock known to be of good quality and of the appropriate genotype for the commodity to be produced in the management situation to be used. The obvious first decision is to choose meat type for meat production and an egg type for egg production. However, having made that decision, it is then necessary to analyse the management situation and market to select a genotype that suits the management situation and produces a commodity suitable for that market. A good example is that of brown eggshells. If the market requires eggs to have brown shells, the genotype selected must be a brown shell layer. Another example would be to choose a genotype best suited for use in a tropical environment. The manager must know in detail the requirements of the situation and then select a genotype best suited to that situation.

Good Husbandry

The following are of major importance when considering the health, welfare and husbandry requirements for a flock:

Confine the Birds

Confining the birds provides a number of advantages:

- Provides a degree of protection from predators.

- Reduces the labour costs in the management of the birds.

- Increases the number of birds that can be maintained by the same labour force.

- Reduces the costs of production.

- Better organisation of the stocking program.

- Better organisation management to suit the type and age of the birds housed.

Importantly, the confinement of the birds at higher stocking densities has a number of disadvantages also including:

- Increases the risk of infectious disease passing from one bird to another.

- Increases the probability that undesirable behavioral changes may occur.

- Increases the probability of a significant drop in performance.

- Birds housed at very high densities can often attract adverse comments.

Protection from a Harsh Environment

A harsh environment is defined as the one that is outside of the comfort range of the birds. In this context high and low temperature, high humidity in some circumstances, excessively strong wind, inadequate ventilation and air movement and high levels of harmful air pollutants such as ammonia are examples of a harsh environment. Much effort is made in designing and building poultry houses that will permit the regulation of the environment to a significant degree.

It is the responsibility of those in charge, and responsible for, the day-to-day management of the birds that the environment control systems are operated as efficiently as possible. To this end, those responsible require a good knowledge of the different factors that constitute the environment and how they interact with each other to produce the actual conditions in the house and, more importantly, what can be done to improve the house environment.

Welfare Needs

A successful poultry house has to satisfy the welfare needs of the birds which vary with the class, age and housing system. Failure to satisfy these needs will, in many cases, result in lower performance from the birds. These needs include:

- The provision of adequate floor space with enough headroom.
- The provision of good quality food with adequate feeding space.
- The provision of good quality water with adequate drinking space.
- The opportunity to associate with flock mates.
- The elimination of anything that may cause injury.
- The elimination of all sources of unnecessary harassment.

Maintenance of Good Health

The presence of disease in the poultry flock is reflected by inferior performance. It is essential that the flock is in good health to achieve their performance potential. There are three elements of good health management of a poultry flock. These are:

- The prevention of disease.
- The early recognition of disease.
- The early treatment of disease.

Prevention of Disease

Preventing the birds from disease is a much more economical way of health management than waiting for the flock to become diseased before taking appropriate action. There are a number of factors that are significant in disease prevention. These are:

- Application of a stringent farm quarantine program:
  - The isolation of the farm/sheds from all other poultry.

- ◦ The control of vehicles and visitors.
- ◦ The introduction of day-old chicks only onto the farm.
- ◦ The prevention of access to the sheds by all wild birds and all other animals including vermin.
- ◦ The provision of shower facilities and clean clothing for staff and visitors.
- ◦ The control of the movement of staff and equipment around the farm.
- • The use of good hygiene practices:
  - ◦ The provision of wash facilities for staff, essential visitors and vehicles prior to entry.
  - ◦ The use of disinfectant foot baths at the entry to each shed.
  - ◦ The thorough cleaning and disinfection of all sheds between flocks.
  - ◦ Maintaining the flock in a good state of well being by good stockmanship, nutrition and housing.
  - ◦ The use of a suitable vaccination program.
  - ◦ The use of a preventive medication program.
  - ◦ The use of monitoring procedures to keep a check on the disease organism status of the farm, to check on the effectiveness of cleaning and sanitation procedures and to test the immunity levels to certain diseases in the stock to check the effectiveness of the vaccination program.

Early Recognition of Disease

Early recognition of disease is one of the first skills that should be learned by the poultry flock manager. Frequent inspection of the flock to monitor for signs of sickness are required. It is expected that inspection of all the birds is the first task performed each day, to monitor for signs of ill health, injury and harassment. At the same time feeders, drinkers and other equipment can be checked for serviceability. If a problem has developed since the last inspection, appropriate action can be taken in a timely manner.

Early Treatment of Disease

If a disease should infect a flock, early treatment may mean the difference between a mild outbreak and a more serious one. It is important that the correct treatment be used as soon as possible. This can only be achieved when the correct diagnosis has been made at an early stage. While there are times when appropriate treatment can be recommended as a result of a field diagnosis i.e. a farm autopsy, it is best if all such diagnoses be supported by a laboratory examination to confirm the field diagnosis as well as to ensure that other conditions are not also involved. When treating stock, it is important that the treatment be administered correctly and at the recommended concentration or dose rate. Always read the instructions carefully and follow them. Most treatments should be administered under the guidance of the regular flock veterinarian.

Nutrition for Economic Performance

Diets may be formulated for each class of stock under various conditions of management, environment and production level. The diet specification to be used to obtain economic performance in any given situation will depend on the factors such as:

- The cost of the mixed diet.

- The commodity prices i.e. the income.

- The availability, price and quality of the different ingredients.

Maximising production is not necessarily the most profitable strategy to use as the additional cost required to provide the diet that will give maximum production may be greater than the value of the increase in production gained. A lower quality diet, while resulting in lower production may bring in greatest profit in the long term because of the significantly lower feed costs. Also the food given to a flock must be appropriate for that class of stock – good quality feed for one class of bird will quite likely be unsuitable for another.

The following are key aspects in relation to the provision of a quality diet:

- The ingredients from which the diet is made must be of good quality.

- The weighing or measuring of all the ingredients must be accurate.

- All of the specified ingredients must be included. If one e.g. a grain is unavailable, the diet should be re-formulated. One ingredient is not usually a substitute for another without re-formulation.

- The micro-ingredients such as the amino acids, vitamins, minerals and other similar materials should not be too old and should be stored in cool storage – many such ingredients lose their potency over time, and particularly so at high temperatures.

- Do not use mouldy ingredients – these should be discarded. Mould in poultry food may contain toxins that may affect the birds.

- Do not use feed that is too old or has become mouldy. Storage facilities such as silos should be cleaned frequently to prevent the accumulation of mouldy material.

Practice of Good Stockpersonship

The term "stockpersonship" is difficult to define because it often means different things to different people. However, "stockpersonship" may be defined as 'the harmonious interaction between the stock and the person responsible for their daily care'. There is no doubt that some stock people are able to obtain much better performance than others, under identical conditions. The basis of good stockpersonship is having a positive attitude and knowledge of the needs and behavior of the stock under different circumstances, of management techniques and a willingness to spend time with the stock to be able to react to any adverse situations as they develop to keep stress to a minimum. Having the right attitude is also a very important element. The stockperson who spends as much time as possible with the stock from day old onward by moving among them, handling them and talking to them, will grow a much quieter bird that reacts less to harassment, is more resistant to disease and performs better.

Maximum use of Management Techniques

There are a number of different management techniques available for use by stockpersons that, while not essential for the welfare of the stock, do result in better performance. Examples of these are the regulation of day length, the management of live weight for age and of flock uniformity. The good manager will utilise these techniques whenever possible to maximise production efficiency and hence profitability of the flock.

Use of Records

There are two types of records that need be kept on a poultry enterprise:

- Those required for financial management – for business and taxation reasons.

- Those required for the efficient physical management of the enterprise.

For records to be of use in the management of the enterprise, they must be complete, current and accurate, be analysed and then used in the decision making process. Failure to use them means that all of the effort to gather the information will have been wasted and performance not monitored. As a result, many problems that could have been fixed before they cause irreparable harm may not be identified until too late.

Marketing

There are three important elements to good marketing practice:

- Produce the commodity required by the consumer: This usually means continuous market research must be carried out to relate production to demand.

- Be competitive: Higher price is usually associated with good quality and specialised product. Therefore, it is necessary to relate price to quality and market demand and to operate in a competitive manner with the opposition.

- Reliability: Produce a commodity for the market and ensure that supply, price and quality are reliable.

POULTRY DISEASES AND MANAGEMENT

Newcastle Disease (ND)/Ranikhet Disease (RD)

Nature of Disease

This is an acute viral disease of poultry characterized by involvement of respiratory system, drop in egg production and mortality as high as 100% in severe cases. This virus has zoonotic effect and can causes human deaths.

Depression

Diarrhoea

Mortality

Causes

Contamination of water

Infected chicken

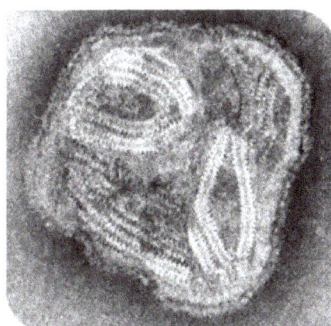
Paramyxovirus

- Paramyxovirus type1 (PMV-1 belongs to the genus Avulavirus, family Paramyxoviridae).

- Based on the disease produced in chickens, NDVs have been classified into five pathotypes, namely Viscerotropic velogenic (most pathogenic), Neurotropic velogenic, Mesogenic (moderate pathogenic), Lentogenic (low pathogenic) and asymptomatic.

- Various physical (heat, irradiation and pH effects) and chemical compounds (potassium permanganate, formalin, ethanol, etc.) could destroy the virus.

- Fumigation of buildings and incubators can be used to kill the virus.

- Infected chickens are the primary source of virus.

- Virus is shed during incubation, during the clinical stage, and for a varying but limited period during convalescence.

- Virus may also be present in eggs laid during clinical disease and in all parts of the carcass during acute virulent infections.

- Infected birds shed virus in exhaled air, respiratory discharges, and feces, etc. which contaminate feed and water.

- Infection in birds occurs through inhalation and ingestion of contaminated materials.

- Wild and pet birds, movement of people and poultry equipment and even poultry products could aid in spread of infection.

- The virus has been found to survive for several days on the mucous membrane of the human respiratory tract and has been isolated from sputum.

Clinical Symptoms

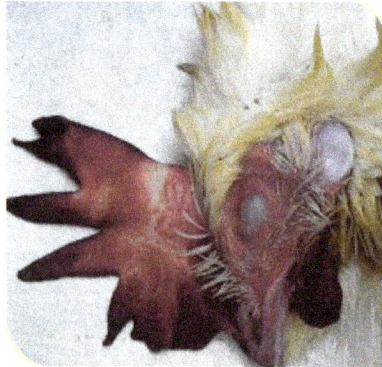

Cyanosis of comb

- Twisting of neck and paralysis of wings and legs,

- Cyanosis of comb,

- Facial edema,

- Diarrhoea,

- Drop in egg production,

- Sudden death.

Gross Lesions

- Haemorrhage in intestine,

- Petechial haemorrhage in proventiculus,

- Congestion and mucoid exudates seen in the respiratory tract, especially in trachea.

Prevention and Control

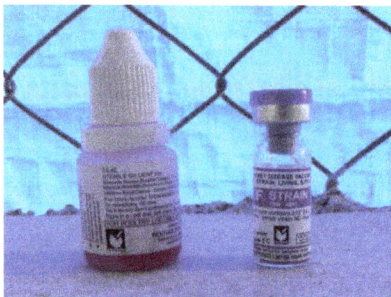

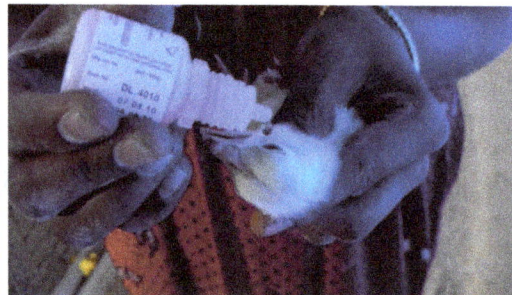

- Disease can be prevented effectively by an integrated approach of vaccination, proper management and strict biosecurity.

- Live virus vaccines both from lentogenic (La Sota, F, B1) and mesogenic (H, R2B, Muktesh-war) strains are used for induction of good immune response.

- Healthy chicks are vaccinated as early as day 1-4 of life.

- Lentogenic strains are administered through ocular (eye) or nasal (nostril) route for primary vaccination of birds.

- Mesogenic strains are administered through subcutaneous or intramuscular route at 6-8 weeks of age as secondary vaccine for better protection of longer duration.

- Killed vaccine with oil adjuvant are used in endemic areas and administered intramuscularly/subcutaneously to maintain high and prolonged antibody titre in layers and breeders and better maternal antibodies in chicks.

Table: Recommended vaccination schedule for layers.

Age in days	Name of the vaccine	Route
5	F/B	I/o (or) I/n
27	LaSota	water
52	LaSota	Water
64	R2B	I/m
112	LaSota	water
280	LaSota	water

Regular disinfection of farm premises and equipment with potassium permanganate (1: 1000), sodium hydroxide (2%) or Lysol (1: 5,000) are useful in preventing this disease.

Infectious Bursal Disease

Nature of Disease

Young chicks upto 0-6 weeks are more susceptible.

This is acute and highly contagious infection of chickens. It is otherwise called as Gumboro disease or Infectious Bursitis or Avian nephrosis. Young chicks upto 0-6 weeks are more susceptible. Morbidity is 100% and mortality is 80-90%. B- Lymphocytes are the primary target cells. It primarily affects the bursa of fabricius, an important organ responsible for immunity. Incubation period is short and clinical signs observed in 2-3 days following infection. Economically significant, because heavy mortality in 3 – 6 wks old chickens and older and severe prolonged immunosuppression of chickens infected at an early age. This disease breakdowns the immunity, leading to the outbreak of other diseases. Immunosuppression leads to vaccination failures, Escherichia coli infection, and Gangrenous dermatitis and Inclusion Body hepatitis – anaemia syndrome.

Causes

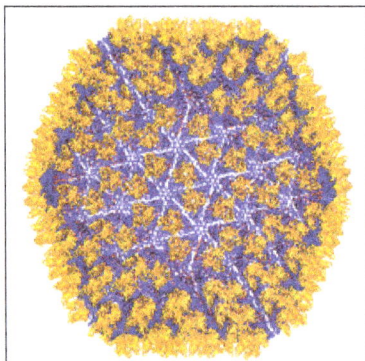

Birna virus

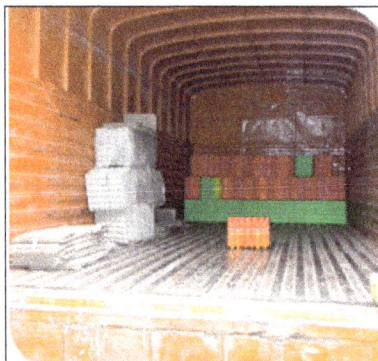

Egg trays, vehicles used in the transport of birds and eggs.

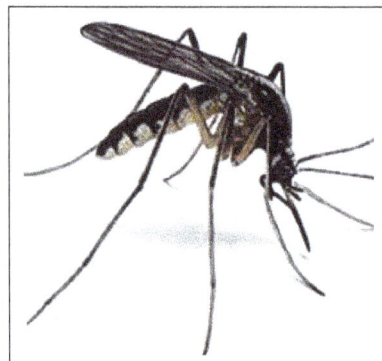

Mosquito

The Birna virus belongs to the family Birna viridae and genus Avibirna virus. This virus is highly contagious and persistent in the environment of poultry houses. Affected birds excrete the virus in faeces for 10-14 days. Virus survives upto 120 days in poultry sheds. Water, feed, droppings from infected birds are viable for 52 days in the poultry houses. Hardy nature of this virus survives heat, cleaning and disinfectant procedures. Survives in the environment between outbreaks. Meal worm, Aedes vexan (mosquito) and litter mites appear to act as carriers and remains infective for up to 8 weeks. Egg trays, vehicles used in the transport of birds, eggs and personal handling of birds in sheds and elsewhere are very important source of carriers of infection. Role of mechanical vectors (human, wild birds, insects). No vertical transmission and carriers. (Disease is not transmitted through eggs). Older birds (due to Bursal regression) are more resistant to infection.

Clinical Symptoms

Closed eyes and death.

Ruffled feathers.

Watery and whitish diarrhoea.

- Self vent pecking,

- Anorexia,

- Depression and trembling,

- Watery and whitish diarrhoea,

- Soiled vents,

- Ruffled feathers,

- Reluctant to move,

- Closed eyes and death.

Gross Lesions

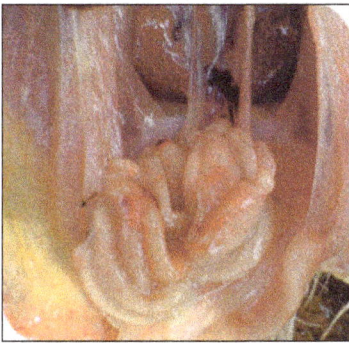

Bursal haemorrhages and enlargement.

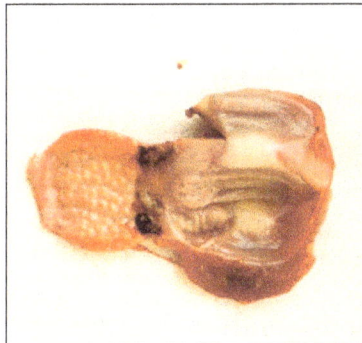

Haemorrhages in proventriculus and Gizzard junction.

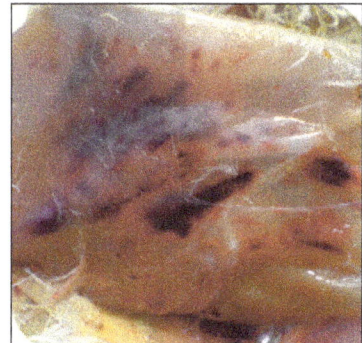

Musle haemorrhages.

- Dehydration of carcass.

- Petechial/paint brush haemorrhages on the leg, thigh and pectoral muscles.

- Hemorrhage in the Proventriculus and Gizzard junction.

- Enlargement of bursa fabricius to almost double its normal size.

- Haemorrhage on the internal and serosal surfaces of the bursa fabricius.

- Intestine with excess mucus.

Prevention and Control

Disposables - Deep burial with slaked lime.

Toxin free feed.

- Primary vaccination with mild or intermediate strain at 2 weeks of age.

- Booster vaccination with intermediate strain (live) after 3 weeks of age.

Table: Recommended vaccination schedule for layer chicks.

Age in days	Name of the vaccine	Route
12-14	IBD Live (Primary)	I/O
22-24	IBD Live (Booster)	I/O

- Vaccination of breeder stock and seromonitoring by hatcheries to ensure adequate levels of maternal antibodies in the chicks.

- To obtain high levels of MDA in progeny, parent stocks are vaccinated between 4 and 10 weeks of age with live vaccine and again at approximately 16 weeks with inactivated oil-adjuvant vaccine.

- Include immuno-stimulants like Vitamin E in the feed.

- Give toxin-free feed.

- Disposal of litter, dead birds, used gunny bags, curtains and other disposables by incineration or deep burial with slaked lime.

- Restricting vehicular movements with crates, egg trays and culled birds.

- Treating feeders and waterers with 5% formalin.

- Fumigating new poultry sheds with formalin fumes.

- Restricting personnel to their sheds for work.

Infectious Coryza

Nature of Disease

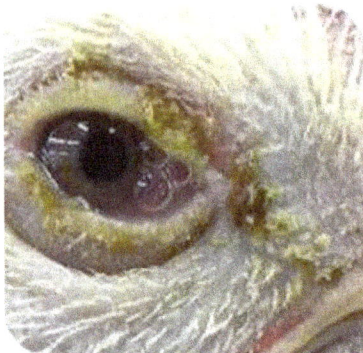

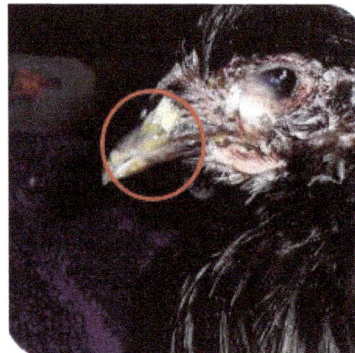

Discharge from the eyes.   Discharge from the nostrils.   Swelling of the face.

- Infectious coryza is an acute, highly contagious, bacterial disease of the upper respiratory tract of chickens.

- A chronic respiratory disease can develop when complicated by other pathogens.

- Characterized by swelling of the face (facial oedema), and discharge from the eyes and nostrils.

Causes

Drinking water contaminated by discharge.

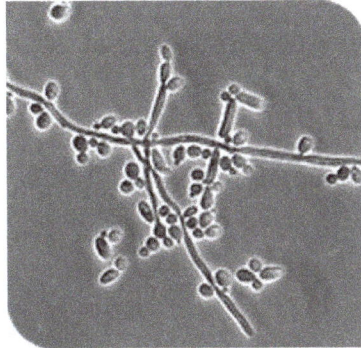
Homophiles paragallinarum.

Older bird suffers more.

- This disease is caused by a bacteria, Homophiles paragallinarum.

- Older bird suffers more severely.

- Clinically affected and carrier birds act as a main source for disease.

- It can be transmitted by drinking water contaminated by nasal discharge as well as by air-borne means over a short distance.

- Lateral transmission occurs readily by direct contact.

- Factors that predispose to more severe and prolonged disease (chronic respiratory disease) include intercurrent infections with microorganisms such as infectious bronchitis virus, Laryngotracheitis virus, Mycoplasma gallisepticum, Escherichia coli or Pasteurella spp. and unfavorable environmental conditions.

- Economic losses are due to marked reduction in egg production (10-40%) in layer.

Clinical Symptoms

Swelling and cyanosis of eyelids and face.

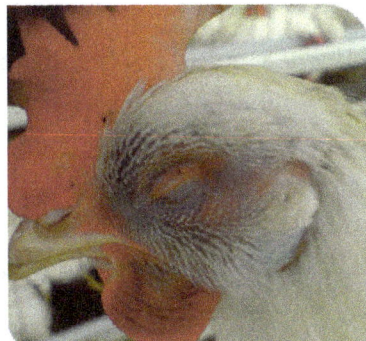
Swelling and cyanosis of eyelids and face.

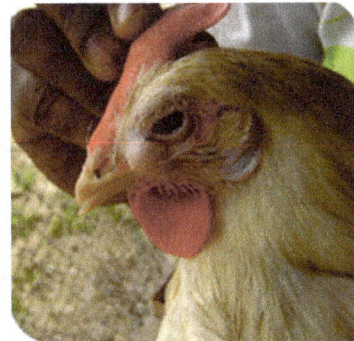
Swelling and cyanosis of eyelids and face.

- The disease in flocks on deep litter management is characterized by rapid spread, high morbidity and low mortality.

- First typical symptoms include sneezing, mucus – like discharge from the opening of the nose, eyes, and swelling on the face (facial oedema).

- In severe case conjunctivitis with closed eyes, swollen wattles, and difficulty in breathing can be seen.

- Feed and water consumption is usually decreased resulting in a drop in egg production.

Gross Lesions

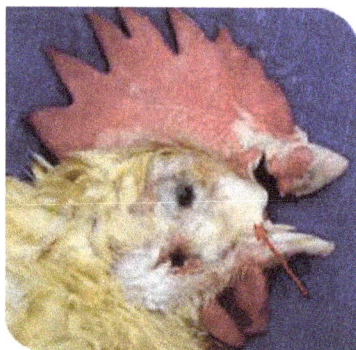

Infraorbital sinus showing consolidated caseous exudate.

Catarrhal to fibrino-purulent inflammation of the nasal passages and infraorbital sinus and conjunctivae. As the disease becomes chronic or other pathogens become involved, the sinus exudates may become consolidated and turn yellowish. Subcutaneous edema of the face and wattles is prominent. The upper trachea may be involved, but the lungs and air sacs are only affected in chronic complicated cases

Prevention and Control

- Prevention of infection into the farm is the best control, which could be achieved by hygiene, sanitation, strict bio security and procurement of birds from disease free sources.

- Since recovered birds are reservoir of infection, such birds should be removed and culled from the flock.

- All–in–all out rearing system is required for eradication of disease.

- Vaccination using inactivated whole culture of organisms containing an adjuvant can protect chickens against the disease.

- In endemic areas, two doses of vaccine, each of which must consist of at least 108 colony-forming units are advocated, given subcutaneously, the first at about 16 weeks of age. Another one is given at 20 weeks of age.

- After cleaning, disinfection and resting of the building for at least 1 week, new birds may be introduced.

- Only day-old chickens or older birds which are known to be free from H. paragallinarum should be used for the restocking.

Infectious Bronchitis

Nature of Disease

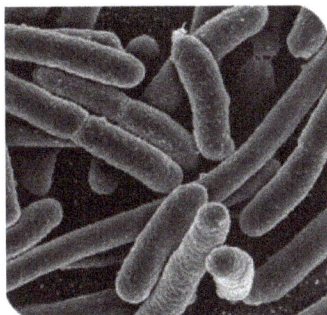
E. coli increases the severity of disease.

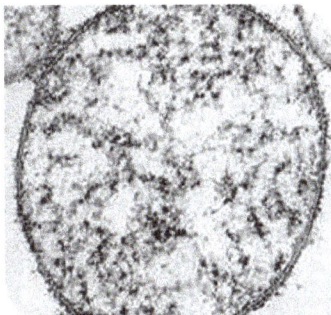
Mycoplasma - increases the severity of disease.

Under 6 weeks age of chicks are more susceptible.

This is a highly infectious and contagious respiratory disease of chicks. Also affect the oviduct, and some strains have a tendency for the kidneys. Disease can occur at any stage, but young chicks, especially under 6 weeks of age are more susceptible. Great economic importance due to its adverse effect on egg production and egg quality in layers, and on production in broilers. Other pathogen such as Mycoplasma or E. coli increases the severity and duration of the disease.

Causes

Contaminated feed.

Egg shells of infected birds.

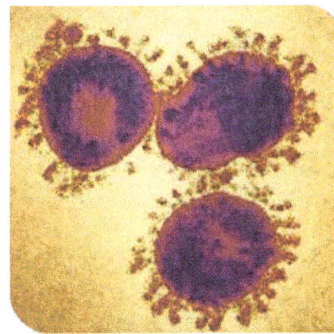
Infectious bronchitis virus.

- The virus belongs to corona group of virus.

- Virus is fragile in nature and gets destroyed by common physical and chemical agents.

- Infection is by inhalation of droplets, through ingestion of feed and water contaminated with virus and by contact with infected birds, contaminated movable equipments, clothing and personnel.

- Virus is present in respiratory discharges, faeces and eggshells of infected birds.

- This virus survives well outside the body during winter therefore, disease incidence and spread is more during winter than other seasons, though disease can occur in any season.

- Spreads very rapidly in the flock.

- Bird to bird by direct transmission.

- Transmission through eggs.

- Fomites also can transmit the disease.

Clinical Symptoms

Misshapen egg with ridges.

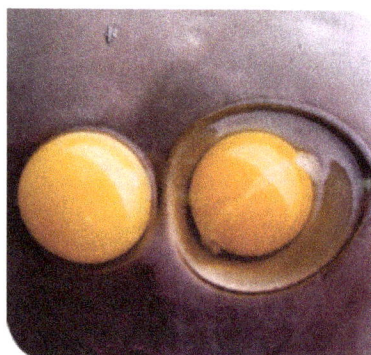
Watery albumin of egg.

- In young chicks upto 6 weeks of age, respiratory signs like sneezing, coughing, gasping, tracheal rales, lachrymation and nasal discharge are more common.

- The chicks will huddle under the hover.

- There may be swelling of sinuses and face.

- In chicks, mortality may as high as 25%-60% and the course of disease is 1-2 weeks.

- Respiratory noises can be heard more distinctly during night when chicks are normally quite.

- In growers and adult birds, signs of less intensity are seen with less occurrence of nasal discharge and mortality is negligible.

- In laying birds, egg production declines (5-50%) rapidly.

- Damage to functional oviduct in adults (most common).

- Egg abnormalities (Production of misshapen, thin or soft –shelled, rough, smaller, corrugations and leathery eggs).

- The egg quality is poor with thin or watery white albumin.

- In uraemic form, birds exhibit depression, ruffled feathers, wet droppings, increased water intake and increased mortality (0.5-1% per week) due to urolithiasis (Kidney stones).

Gross Lesions

Serous, catarrhal, or caseous exudates in the trachea and bronchi lumen, generally without haemorrhages. Plugs of yellow caseous material obstructing the bronchi and lower parts of trachea of chicks that dies. Fluid yolk material may be found in the abdominal. Abnormal ovary having the misshapen follicle. Middle third of the oviduct may appear atrophied and ova ruptured into abdominal cavity. Swollen, pale kidneys and deposits of urates in kidney, ureters and throughout the body.

Prevention and Control

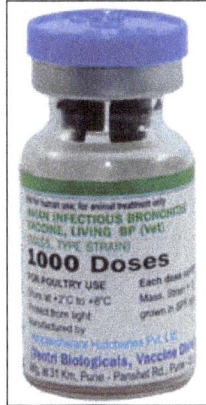

Strict hygienic management procedures and vaccination can prevent the disease.

Table: Recommended vaccination schedule for layers.

Age in days	Name of the vaccine	Route
5-7	IB Live	I/O
28-30	IB Live	I/O
80	IB Live	D/W
112-114	IB Killed	S/C
280	IB Killed	S/C

Gout

Nature of Disease

Death - due to kidney failure.   Deposition of urates on joints.   Laying hens fed high level of calcium.

- Gout is a not disease condition, but a clinical sings of severe kidney dysfunction.

- Characterized by presence of high level of uric acid in the blood.

- Deposition of urates on the surface of various internal organs or joints (especially hock joint).

- Death is due to kidney failure.

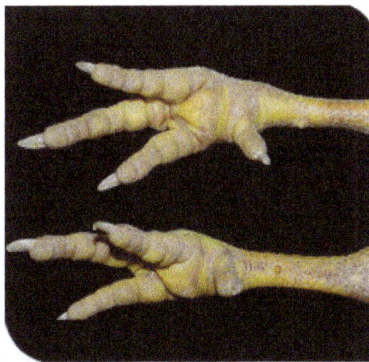

- It is a main problem of laying hens fed high level of calcium.

- Two distinct forms are there visceral gout and articular gout.

Causes

Dehydration.

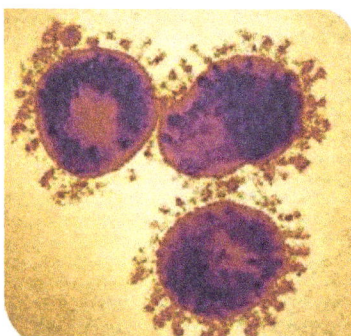
Infectious bronchitis virus.

Treatment with sodium bicarbonate.

- Kidney dysfunction leads to hyperuricaemia,

- Dehydration,

- Excessive dietary calcium or calcium: phosphorus imbalance,

- Vitamin A deficiency,

- Increased intake of protein,

- Intake of excessive amount of salt,

- Infection with infectious bronchitis virus in young chicken,

- Urolithiasis and mycotoxins,

- An electrolyte excess or deficiency,

- Prolonged treatment with sodium bicarbonate.

Clinical Symptoms

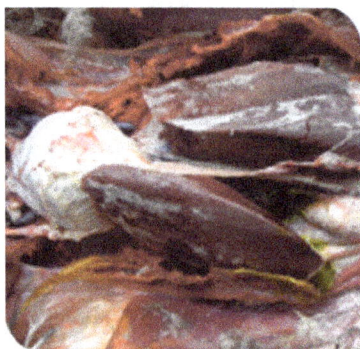
Deposition of urate salts as a white chalky coating on organs.

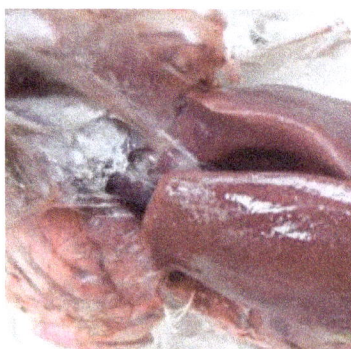
Deposition of urate salts as a white chalky coating on organs.

Affected leg joints.

- In articular gout joints are much swollen, with deposition of masses of chalk-like material. Usually wing and leg joints are affected.

- Affected bird cannot move and so die of starvation.

- Articular gout occurs mostly in male birds but visceral gout occurs both in male and female.

Gross Lesions

In articular gout tissue surrounding the joints is white due to urate deposition. In Visceral gout kidneys are swollen and congested and greyish white in color. Apart from kidney, chalk-like crystals are deposited on the serous membranes of various internal organs like mesentery, peritoneum, heart, proventriculus and lungs. Urate deposition appears as a white chalky coating on organs.

Prevention and Control

Increase maize.

Water containing electrolytes.

- Avoid feeding of high level of calcium in advance of sexual maturity.

- Reduce high level of protein.

- Increase maize, and formulate the feed.

- Give plenty of water containing electrolytes.

Fowl Pox

Nature of Disease

Disease affects birds of all ages.

poor weight gain.

Fowl pox is a common, slow spreading, widely prevalent, contagious viral disease of poultry. This is characterized by wart like growth over the non-feathered parts of skin and mucosa of the upper respiratory and digestive systems. Disease affects birds of all ages. Economically very important as it can cause poor weight gain, drop in egg production in layers and mortality. It has no zoonotic importance.

Causes

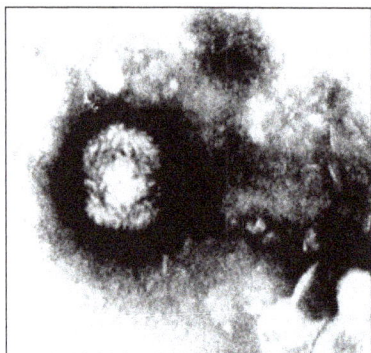
Avipoxvirus.

Overcrowding.

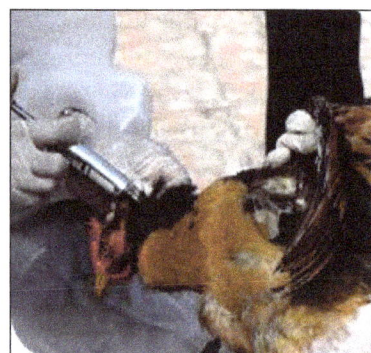
Through vaccination virus transfer from one to another.

- Disease is caused by an avipoxvirus, which is relatively resistant to common disinfecting procedures and can survive in dried scab for years.

- Virus does not penetrate intact skin. Some break in the skin is required for the virus to enter the epithelial cells, replicate and cause disease.

- Disease spreads mechanically through direct contact, where contaminated materials soil the abraded/lacerated skin, by mosquitoes and other biting insects.

- In a contaminated environment, aerosols (droplets) generated from feather follicles and dried scabs carry the virus and spread the disease.

- It can be transmitted by the respiratory tract.

- During vaccination, the individuals may transfer the virus from affected to healthy birds.

- Deposition of virus in eyes, through lachrymal duct it goes to larynx and causes upper respiratory tract infection.

- Disease is frequently observed during rainy and winter seasons with persisting overcrowding and unhygienic conditions.

- Disease remains a problem for a long time, in farms where multiple age birds are maintained, even after preventive vaccinations.

Clinical Symptoms

- This disease is manifested clinically in one of the two forms, cutaneous or diphtheritic or both.

- Poor weight gain and drop in egg production in layers is always significant.

- In cutaneous form (dry pox), lesions in the form of scab appears on the comb, wattles, eyelids, external nares, corner of the beak and other non-feathered parts of the body.

- Lesions on eye may affect the bird's vision or close both eyes affecting the ability of the bird to reach feed and water leading to starvation and deaths.

- Papule —> vesicles —> pustules —> crust/scab —> scar formation.

- In cutaneous form, flock mortality is usually low rarely exceeding 25%.

- In diphtheritic form (wet pox or fowl diphtheria), initially small nodules are formed on the mucous membrane of mouth, oesophagus and trachea.

- Later on, these nodules become yellow cheesy in nature, thereby, forming a diphtheritic membrane on these organs causing obstruction, interference with feeding and difficulty in breathing with mortality up to 50%.

Wart like gowth in the non-feathered parts of the body.

Wart like gowth in the non-feathered parts of the body.

Wart like gowth in the non-feathered parts of the body.

Gross Lesions

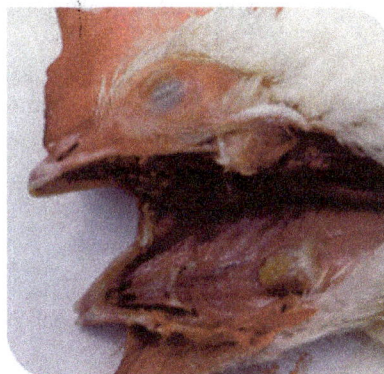

Fowl pox gross.

- Lesions initially start as nodular area with blanched appearance (papule).

- It becomes enlarged and yellowish (pustules) terminating into thick, dark scab.

Prevention and Control

- Two types of vaccines (pigeon pox and fowl pox vaccines) can be used for vaccination.

- Pigeon pox vaccine is less pathogenic and can be used on chickens at any stage by wing web method and it produces immunity for 6 months therefore revaccination is required.

- Fowl pox vaccine produces solid immunity, usually carried out at 6-8 weeks of age by intra-muscular route or wing web method.

- Successful or effective vaccination can be judged by takes, where examination of vaccinated birds after 7-10 days of vaccination, show swelling or a scab at the site of puncture or vaccine application.

- Absence of takes indicates poor potency of the vaccine, presence of maternal antibodies and improper vaccination.

- In such cases, revaccination with new batch/lot of vaccine should be done.

- During disease outbreak, affected birds if less than 30% should be segregated immediately and the remaining birds must be vaccinated at earliest possible.

- Standard sanitation and strict biosecurity measures can control the fowl pox.

- Disinfection of premises with sodium hydroxide (1:500), cresol (1:400) and phenol (3%) proved beneficial in control of fowl pox.

Disinfection of premises.

Sanitation.

Colibacillosis

Nature of Disease

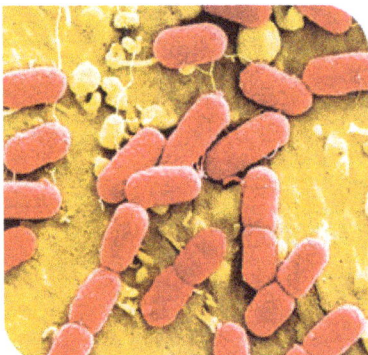

Avian pathogenic Escherichia coli.

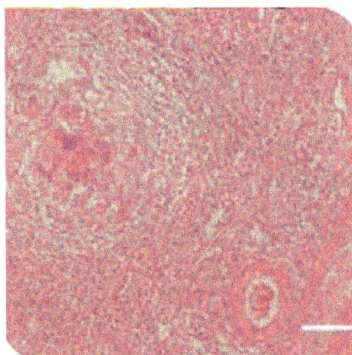

Coligranuloma.

Mushy chick disease.

Colibacillosis is a localized or systemic infection caused by avian pathogenic Escherichia coli bacteria characterized by septicemia, drop in production and mortality. Colisepticaemia, egg peritonitis, yolk sac infection("mushy chick disease" and "omphalitis'), and coligranuloma (Hjarre's disease) are the well recognized results of E. coli infection. These conditions are collectively grouped under the heading "Colibacillosis".

Causes

| Birds consuming beetles. | Birds in poor environmental condition. | Contaminated eggs. |

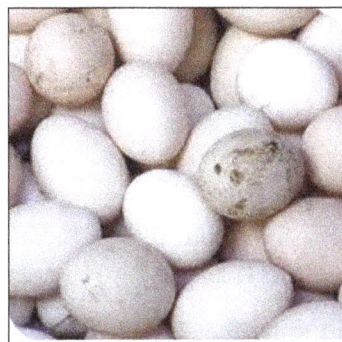

- Escherichia coli is a normal inhabitant of the digestive tract and wild birds, which are source of infection for poultry flocks and spread is by direct or indirect contacts.

- Organisms are susceptible to common physical and chemical disinfectants. They can survive in freezing and can remain viable for long period at low temperatures.

- In litter, high ammonia concentration can inactivate rapidly and survive at 37 °C for 1-2 days.

- Trachea, caeca and oviduct of recovered birds can harbor E. coli for several weeks.

- Organisms are not transmitted through eggs.

- Beetles can transmit the bacteria and birds consuming these beetles get the infection.

- Large numbers of E. coli are maintained in the poultry house environment through fecal contamination.

- Initial exposure to pathogenic E. coli may occur in the hatchery from infected or contaminated eggs, but systemic infection usually requires predisposing environmental factors or infectious causes.

- Mycoplasmosis, infectious bronchitis, Newcastle disease, hemorrhagic enteritis, and turkey bordetellosis precede colibacillosis.

- Poor air quality and other environmental stresses may also predispose to E. coli infections.

- A hatching environment that is not sufficiently humid is often associated with a high incidence of yolk sac infection. E. coli multiplies rapidly in the intestines of newly hatched chicks and infection spreads rapidly from chick to chick in the hatchery and brooders.

Clinical Symptoms

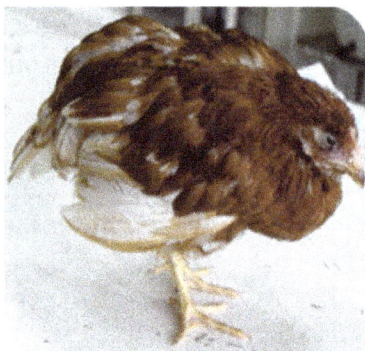
Depression and disinclination to move.

Mortality in a batch of chicks - first week after hatching.

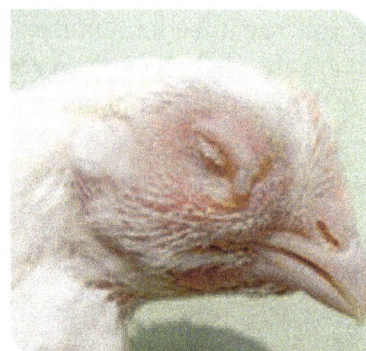
Swollen Head Syndrome.

- Affected birds will have respiratory symptoms, depression, loss of appetite and disinclination to move.

- Soiling of vent with pasty diarrhoeic faeces will be seen.

- In Coliform omphalitis/yolk sac infection causes mortality in a batch of chicks in the first week of life after hatching. Affected chicks exhibit depression, sleepiness with tendency to huddle together around heat sources with distended abdomen and swelling of naval.

- Swollen Head Syndrome (SHS) causes swelling of head.

Gross Lesions

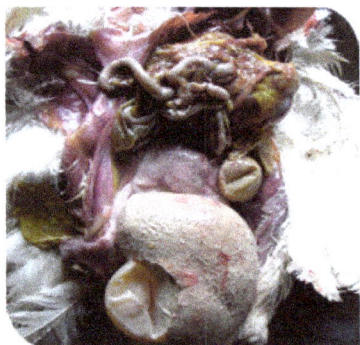
Cassiated egg mass inside the oviduct.

Egg peritonitis.

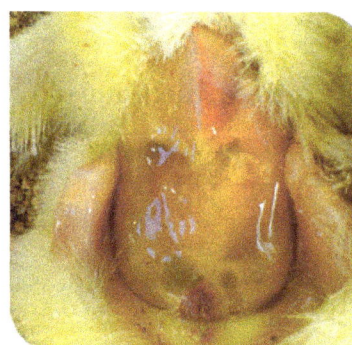
Inflammed unabsorbed yolk sac with abnormal colour.

Birds with Colisepticaemia develop airsacculitis, perihepatitis and pericarditis with cloudiness of pericardial sac and light coloured fibrinous exudates. In Coliform omphalitis/yolk sac infection, unabsorbed yolk will have abnormal volume, colour, consistency and smell. Salphingitis (inflammation of oviduct) causes peritonitis, abdominal cavity containing abnormal egg or yolk mass, impaction of oviduct, distortion of ovaries and produce foul smelling pus like material. Coligranuloma (Hjarre's disease) causes millet sized multiple projected granulomas are found on the surface of liver, caeca and mesentery. Air sac disease (Chronic Respiratory Disease) causes thickened air sac and will have caseous exudates. A whole or partly formed egg may be seen in the abdominal cavity known as impaction of oviduct which was firm and adhered to peritoneum and visceral organs.

Prevention and Control

Biosecurity

- Treatment strategies include attempts to control predisposing infections or environmental factors and early use of antibacterials indicated by susceptibility tests.

- Maintain adequate hygiene and bio security to minimize transmission through faecal contamination of eggs.

- Infected excreta and litter should be disposed properly to avoid contamination of natural water sources and spread in the farm.

- Avoid stress and overcrowding in the flocks which favors outbreaks.

- Birds should be procured from sources tested free of ND, IB, and Mycoplasma.

- Diet with protein, selenium, and vitamin E may be favorable in control of colibacillosis.

- Chlorination of drinking water inactivates the bacteria.

Coccidiosis

Nature of Disease

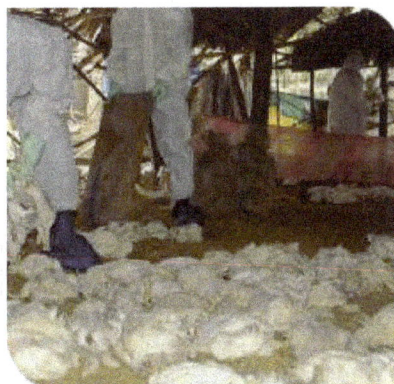
Heavy mortality in broilers.

Coccidiosis is one of the most important protozoan diseases of poultry. Outbreaks are common between 3- 6 weeks of age. Inflict heavy mortality in broilers and also in growers raised on deep litter. One of the biggest causes of economic losses to poultry.

Causes

Rodent

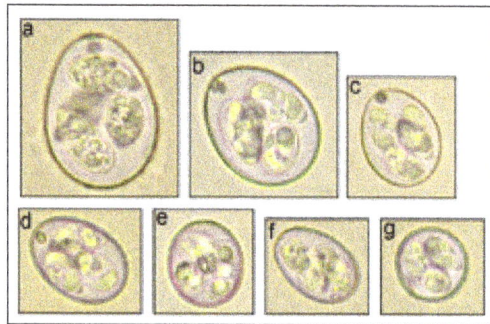
Seven species of genus Eimeria.

- It is caused by seven species of genus Eimeria.

- There is no cross immunity (cross-protection) among the seven species of Eimeria.

- Of the seven disease-producing species, E. tenella, E. necatrixand E. brunettiare the most harmful and cause high morbidity and high mortality.

- E. tenellaaffects caeca and E. necatrixmiddle portion and E. brunettilower portion of the small intestine.

- E. maximaand E. acervulinaare moderately harmful. E. maximaaffects middle portion of the small intestine, and E. acervulinamainly upper portion, that is, duodenum.

- Ingestion of the infective form of oocysts (sporulated oocysts) is the only natural method of spread.

- Ingestion of feed and water contaminated with sporulated oocysts causes the infection.

- Infected chickens may shed oocysts in the faeces for several days or weeks.

- Oocysts can be spread mechanically by movement of people, equipment and foot wear between farms.

- It can also spread through cockroaches, rodents, pets and wild birds.

Clinical Symptoms

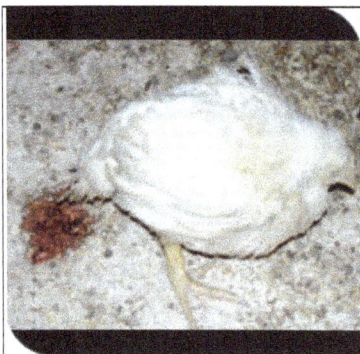
Bloody diarrhoea.

Dehydration.

Drooping wings.

- Affected birds appear dehydrated with drooping wings and ruffled feathers.

- They may huddle together and severe watery or bloody diarrhoea.

- Droppings of affected birds usually contain blood, fluid, and mucus.

- Emaciated and anaemic appearance.

- High mortality and most occurs between 5 and 6 days following infection.

- Causes poor weight gain.

- Egg production may be reduced in laying birds.

Gross Lesions

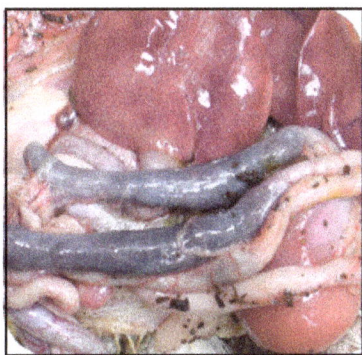

Distended small intestine filled with fluids and clotted blood.

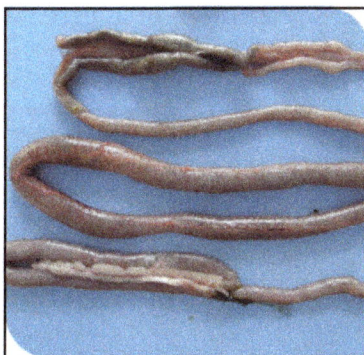

Distended small intestine filled with fluids and clotted blood.

Distended small intestine filled with fluids and clotted blood.

- Caeca enlarged with clotted blood.

- Middle portion of the small intestine is distended to twice its normal size.

- Intestinal lumen filled with blood.

- Lining of the small intestine covered with tiny hemorrhages.

- Intestinal mucosa swollen and thickened.

Prevention and Control

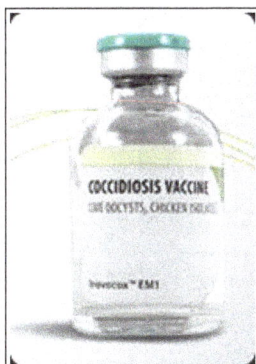

Coccidiosis Vaccine.

Use lime powder to dry the litters.

Use slphonamide as feed additive.

- Coccidiosis is far more easily prevented than treated.

- Control depends mainly on drugs, although an effective vaccine is now available for breeder or layer replacements.

- Two types of vaccines have been used to obtain immunity (protection) against coccidiosis.

- Birds are vaccinated through drinking water between the age of 5 and 9 days.

- The use of one anticoccidial in the starter and another in the grower feed is called a 'shuttle programme'.

- The use of shuttle programme has been found to reduce drug resistance.

- Hygiene and biosecurity along with measures to prevent contamination of feed and water with droppings can prevent the occurrence of infection.

- Dry litter and raking of litter at regular intervals prevent sporulation of oocysts, thereby minimizes chances of infection.

- During rainy season lime powder can be used to dry the litter and to render the oocysts ineffective.

- Use of anticoccidial in feed as feed additives (e.g. Ionophores, sulphonamides, and quinolones) can avoid acute outbreak.

- Rotation of anticoccidial every 4-6 months are helpful in maintaining their efficacy and restricting the development of resistance.

- In disease condition, medication must be given in consultation with qualified veterinarian to avoid complications.

Ascariasis

Nature of Disease

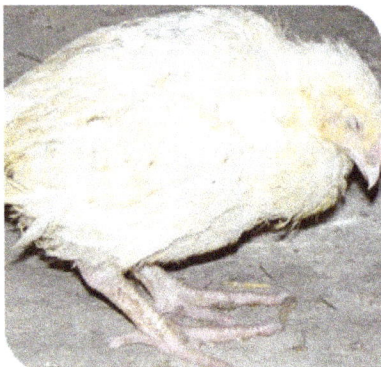

Anaemia and weight loss.

Young birds up to 3 months of age are more susceptible.

- This is a worm infestation of chickens.

- Young birds up to 3 months of age are more susceptible.

- It has significant effect on the production of birds.

- It causes diarrhoea, anaemia, weight loss, increased mortality and reduced egg production.

Causes

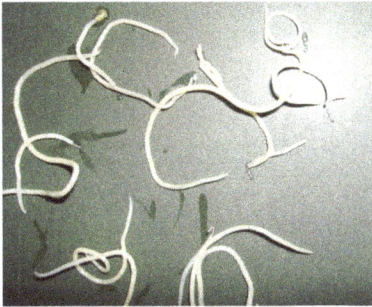
Ascaridia galli.

Malnourished bird.

Reusing of litter materials.

- Ascaridia galli is the species found in intestines of chickens.

- Ascaridia are the largest roundworm of birds.

- Parasites usually affect malnourished flocks.

- Serious infection occurs if the litter is reused in the case of broilers.

- Dietary deficiencies such as vitamin A, B and B12, various minerals and proteins leads to heavy infection.

- A. galli eggs are ingested by grass hoppers or earth worms, hatch, and are infective to chickens.

- Under optimum conditions of temperature and moisture, eggs in the droppings become infective in 10 – 12 days.

- Eggs are quite resistant to low temperatures.

- Chicken over 3 months are more resistant to infection.

Clinical Symptoms

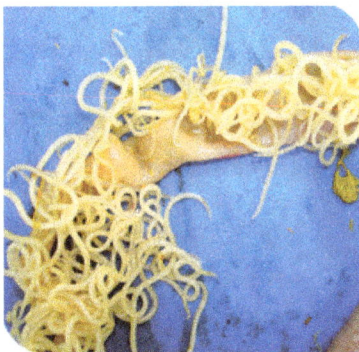
Occlusion of thread like worms in intestine.

Anaemic

Decreased egg production.

- Cause poor bodily condition and weight loss.

- Birds become anaemic and suffer from diarrhoea.

- In heavy affection, it causes haemorrhagic enteritis.

- Loss of blood and increased mortality.

- Affected birds become unthrifty, markedly emaciated and egg production is decreased.

Gross Lesions

- Occlusion of thread like worms in the intestine.

- In heavy infection, intestinal obstruction may occur.

- Ascaridia wormsmay found in the hen's egg. Infected eggs can be detected by candling.

Prevention and Control

Changing litter material

Clean feeding troughs

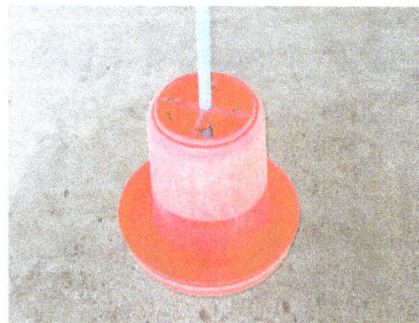
Clean feeding and water

- Regular deworming of chicks with piperazine compounds is highly effective.

- Young birds should be separated from older birds.

- Avoid reuse of litter materials.

- Changing of litter can reduce infections.

- Treatment of the soil or litter to kill intermediate hosts is useful.

- Litter may be treated with suitable insecticides.

- Litter materials should always be kept in dry condition.

- Extreme care should be taken to ensure that feed and water are not contaminated.

- Provide clean feeding troughs and drinking water appliances.

- Poultry runs should be well drained.

Avian Adenovirus

Aviadenoviruses are adenoviruses that affect birds - particularly chickens, ducks, geese, turkeys and pheasants. There are currently eight species in this genus including the type species Fowl aviadenovirus A. Viruses in this genus cause specific disease syndromes such as Quail Bronchitis (QB), Egg Drop Syndrome (EDS), Haemorrhagic Enteritis (HE), Pheasant Marble Spleen Disease (MSD), Falcon adenovirus A and Inclusion Body Hepatitis (IBH). Avian adenoviruses have a worldwide distribution and it is common to find multiple species on a single farm. The most common serogroups are serogroup 1, 2 and 3.

Transmission

No evidence of transmission from birds to humans has been identified. The virus is mainly spread horizontally by the oro-fecal route, but vertical transmission can occur in serogroup 1. Once it has infected the bird the virus may remain latent until a period of stress, when it may then cause clinical disease.

Clinical Signs and Diagnosis

Infections are normally subclinical, however clinical disease can occur - especially in young birds as maternal antibody levels begin to wane.

Clinical signs are related to the organ affected. Signs of gastrointestinal disease (Haemorrhagic Enteritis) include diarrhea, anorexia, melena and hematochezia. Anaemia and dehydration may develop secondary to this haemorrhagic enteritis. Signs of reproductive disease (Egg Drop Syndrome) include low egg production/hatching and the laying of abnormal eggs (size, shape, colour, texture). Adenovirus infection may infect other organs, causing a splenitis, inclusion body hepatitis, bronchitis, pulmonary congestion ventriculitis, pancreatitis, or oedema, depending on the species of bird infected.

Diagnosis of aviadenovirus is by histopathology, electron microscopy, viral isolation, ELISA and PCR. In addition, virus antigen can be detected double immunodiffusion. Postmortem examination may reveal a variety of clinical signs relating directly to the organ affected. Specifically, Egg Drop Syndrome can be diagnosed by hemagglutinin inhibition and the virus causes haemagglutination in chickens and ducks.

Treatment and Control

Vaccines for HE and EDS are available (ATCvet codes: QI01AA05 (WHO) for the inactivated vaccine, QI01AD05 (WHO) for the live vaccine, plus various combinations). Disease incidence may be reduced by minimising stress levels, using appropriate hygiene measures and providing adequate nutrition.

Structure

Viruses in Aviadenovirus are non-enveloped, with icosahedral geometries, and T=25 symmetry. The diameter is around 90 nm. Genomes are linear and non-segmented, around 35-36kb in length and have a guanine/cytosine content of 53–59%. The genome codes for 40 proteins.

Genus	Structure	Symmetry	Capsid	Genomic arrangement	Genomic segmentation
Aviadenovirus	Polyhedral	Pseudo T=50	Non-enveloped	Linear	Monopartite

Life Cycle

Viral replication is nuclear. Entry into the host cell is achieved by attachment of the viral fiber glycoproteins to host receptors, which mediates endocytosis. Replication follows the DNA strand displacement model. DNA-templated transcription, with some alternative splicing mechanism is the method of transcription. The virus exits the host cell by nuclear envelope breakdown, viroporins, and lysis. Birds serve as the natural host.

Genus	Host details	Tissue tropism	Entry details	Release details	Replication site	Assembly site	Transmission
Aviadenovirus	Birds	None	Glycoprotiens	Lysis	Nucleus	Nucleus	Unknown

Avian Infectious Bronchitis

Avian infectious bronchitis (IB) is an acute and highly contagious respiratory disease of chickens. The disease is caused by avian infectious bronchitis virus (IBV), a coronavirus, and characterized by respiratory signs including gasping, coughing, sneezing, tracheal rales, and nasal discharge. In young chickens, severe respiratory distress may occur. In layers, respiratory distress, nephritis, decrease in egg production, and loss of internal (watery egg white) and external (fragile, soft, irregular or rough shells, shell-less) egg quality are reported.

Signs and Symptoms

Coughing and rattling are common, most severe in young, such as broilers, and rapidly spreading in chickens confined or at proximity. Morbidity is 100% in non-vaccinated flocks. Mortality varies according to the virus strain (up to 60% in non-vaccinated flocks). Respiratory signs will subdue within two weeks. However, for some strains, a kidney infection may follow, causing mortality by toxemia. Younger chickens may die of tracheal occlusion by mucus (lower end) or by kidney failure. The infection may prolong in the cecal tonsils.

In laying hens, there can be transient respiratory signs, but mortality may be negligible. However, egg production drops sharply. A great percentage of produced eggs are misshapen and discolored. Many laid eggs have a thin or soft shell and poor albumen (watery), and are not marketable or proper for incubation. Normally-colored eggs, indicative of normal shells for instance in brown chickens, have a normal hatchability.

Egg yield curve may never return to normal. Milder strains may allow normal production after around eight weeks.

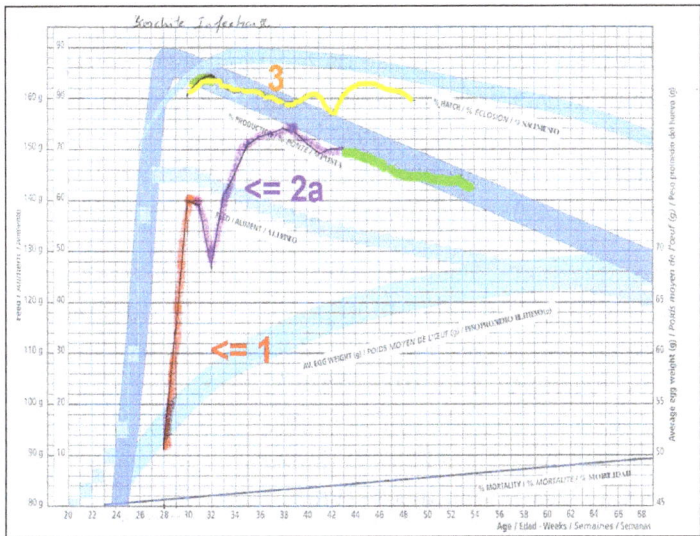

Egg yield curve in BI in a parent flock.

Thin-shelled egg.

Abnormal granulations on shell.

Soft-shelled eggs.

Misshapen and discolored eggs.

Cause

IBV was the first coronavirus described and varies greatly genetically and phenotypically, with hundreds of serotypes and strains described. Coronaviruses contain the largest known viral RNA genome in number of nucleotides, of approximately 30,000 bases. The RNA forms a single strand and single segment. IBV diversity is based on transcriptional error, which may become very relevant if occurring in genomic sequences coding for proteins, involved in adsorption to target cell or inducing immune responses. Transcriptional error variants may emerge with evolutionary advantage in susceptible chickens. Large genomic changes will occur with entire gene interchanges, by reassortment, as for its replication, seven subgenomic mRNAs are produced and will enable reassortment in coinfections.

Diagnosis

Chicken respiratory diseases are difficult to differentiate and may not be diagnosed based on respiratory signs and lesions. Other diseases such as mycoplasmosis by Mycoplasma gallisepticum (chronic respiratory disease), Newcastle disease by mesogenic strains of Newcastle diseases virus (APMV-1), Avian metapneumovirus, infectious laryngotracheitis, avian infectious coryza in some stages may clinically resemble IB. Similar kidney lesions may be caused by different etiologies, including other viruses, such as infectious bursal disease virus (the cause of Gumboro disease) and toxins (for instance ochratoxins of Aspergillus ochraceus), and dehydration.

In laying hens, abnormal and reduced egg production are also observed in Egg Drop Syndrome 76 (EDS), caused by an Atadenovirus and avian metapneumovirus infections. At present, IB is more common and far more spread than EDS. The large genetic and phenotypic diversity of IBV have been resulting in common vaccination failures. In addition, new strains of IBV, not present in commercial vaccines, can cause the disease in IB vaccinated flocks. Attenuated vaccines will revert to virulence by consecutive passage in chickens in densely populated areas, and may reassort with field strains, generating potentially important variants.

Definitive diagnosis relies on viral isolation and characterization. For virus characterization, recent methodology using genomic amplification (PCR) and sequencing of products, will enable very precise description of strains, according to the oligonucleotide primers designed and target gene. Methods for IBV antigens detection may employ labelled antibodies, such as direct immunofluorescence or immunoperoxidase. Antibodies to IBV may be detected by indirect immunofluorescent antibody test, ELISA and Haemagglutination inhibition (haemagglutinating IBV produced after enzymatic treatment by phospholipase C).

Treatment and Prevention

No specific treatment is available, but antibiotics can be used to prevent secondary infections. Vaccines are available (ATCvet codes: QI01AA03 (WHO) for the inactivated vaccine, QI01AD07 (WHO) for the live vaccine; plus various combinations).

Biosecurity protocols including adequate isolation, disinfection are important in controlling the spread of the disease.

Duck Plague

Duck plague is a highly contagious infectious disease of ducks, geese, and swans caused by a herpes virus. This disease was first described in domestic waterfowl in the Netherlands in 1923. Duck plague was first reported in the United States in commercial ducks and wild waterfowl on Long Island, New York in 1967.

Significance

Duck plague outbreaks in domestic waterfowl can result in mortality rates as high as 60% and can cause significant decreases in egg production that can lead to major economic losses. This virus may also cause major die-offs in migratory waterfowl, but there have only been three such known outbreaks involving wild waterfowl in the United States.

Species Affected

Duck plague affects ducks, geese, and swans, though some species are more susceptible than others. Some of the more highly susceptible species include the blue-winged teal, redhead duck, and wood duck. The Canada goose, gadwall, mallard, and Muscovy are moderately susceptible, while the pintail is less susceptible. Coots may also be susceptible to duck plague. Non-waterfowl bird species are believed to be resistant to the virus. Duck plague does not infect mammals, nor does it infect humans.

Transmission

Susceptible birds can become infected with duck plague by coming into contact with a bird that is shedding the virus, or by coming into contact with an environment that is contaminated with feces or oral and nasal secretions from an infected bird. Ingesting food or water contaminated with the virus is a common mode of transmission. Susceptible birds may become infected by way of inhalation of aerosolized secretions containing the virus. The virus can also be passed in eggs from a female to her offspring. In the United States, the majority of outbreaks occur from March to June when birds are crowded during spring migration and under stress due to seasonal weather changes, migration, and breeding. These stressors make the birds more susceptible to disease and the overcrowding facilitates disease transmission.

Clinical Signs

Clinical signs can vary depending on the viral strain, and the species affected, age, sex, and immune status of the infected bird. Clinical signs may develop within 3-7 days of exposure to the virus and include depression, ruffled feathers, difficulty breathing, loss of coordination, avoidance of light, extreme thirst, loss of appetite, ocular and nasal discharge (that may be bloody), bloody and watery diarrhea, and a bloody vent. Females may experience decreased egg production, while males may have a prolapsed penis. Birds die quickly of duck plague, so they are often found dead before clinical signs have been observed. Some birds will survive and become asymptomatic carriers of the virus and can act as sources for future outbreaks. These birds often have a "cold sore" like ulcer under the tongue from which the virus is shed. At necropsy, birds that died of duck plague will often have hemorrhagic or necrotic bands in the intestines and a digestive tract that is filled with blood. There may also be cheese-like plaques within the esophagus and cloaca.

- Diagnosis: Duck plague is diagnosed through laboratory testing by isolating and identifying the virus.

- Treatment: As with most viral diseases, there is no treatment for duck plague.

Management

Prompt carcass disposal and environmental decontamination is key in control of duck plague outbreaks in wild flocks. When outbreaks occur in domestic flocks, the entire flock is often destroyed to prevent surviving birds from carrying the virus and initiating future outbreaks. The premises must also be thoroughly decontaminated and the carcasses burned. Vaccines against duck plague are used in the commercial duck industry, but the vaccines may not be effective in preventing illness in wild waterfowl. Domestic birds should be prevented from coming into contact with wild birds. The United States Fish and Wildlife Service has set up a national monitoring system for duck plague, so that outbreaks in wild and domestic flocks will be recognized and responded to quickly.

Avian Influenza

Avian influenza—known informally as avian flu or bird flu is a variety of influenza caused by viruses adapted to birds. The type with the greatest risk is highly pathogenic avian influenza (HPAI). Bird flu is similar to swine flu, dog flu, horse flu and human flu as an illness caused by strains of influenza viruses that have adapted to a specific host. Out of the three types of influenza viruses (A, B, and C), influenza A virus is a zoonotic infection with a natural reservoir almost entirely in birds. Avian influenza, for most purposes, refers to the influenza A virus.

Though influenza A is adapted to birds, it can also stably adapt and sustain person-to person transmission. Recent influenza research into the genes of the Spanish flu virus shows it to have genes adapted from both human and avian strains. Pigs can also be infected with human, avian, and swine influenza viruses, allowing for mixtures of genes (reassortment) to create a new virus, which can cause an antigenic shift to a new influenza A virus subtype which most people have little to no immune protection against.

Avian influenza strains are divided into two types based on their pathogenicity: high pathogenicity (HP) or low pathogenicity (LP). The most well-known HPAI strain, H5N1, appeared in China in 1996, and also has low pathogenic strains found in North America. Companion birds in captivity are unlikely to contract the virus and there has been no report of a companion bird with avian influenza since 2003. Pigeons can contract avian strains, but rarely become ill and are incapable of transmitting the virus efficiently to humans or other animals.

Between early 2013 and early 2017, 916 lab-confirmed human cases of H7N9 were reported to the World Health Organization (WHO). On 9 January 2017, the National Health and Family Planning Commission of China reported to WHO 106 cases of H7N9 which occurred from late November through late December, including 35 deaths, 2 potential cases of human-to-human transmission, and 80 of these 106 persons stating that they have visited live poultry markets. The cases are reported from Jiangsu (52), Zhejiang (21), Anhui (14), Guangdong (14), Shanghai (2), Fujian (2) and Hunan (1). Similar sudden increases in the number of human cases of H7N9 have occurred in previous years during December and January.

Genetics

Genetic factors in distinguishing between "human flu viruses" and "avian flu viruses" include:

- PB2: (RNA polymerase): Amino acid (or residue) position 627 in the PB2 protein encoded by the PB2 RNA gene. Until H5N1, all known avian influenza viruses had a Glu at position 627, while all human influenza viruses had a Lys.

- HA: (hemagglutinin): Avian influenza HA viruses bind alpha 2-3 sialic acid receptors, while human influenza HA viruses bind alpha 2-6 sialic acid receptors. Swine influenza viruses have the ability to bind both types of sialic acid receptors. Hemagglutinin is the major antigen of the virus against which neutralizing antibodies are produced, and influenza virus epidemics are associated with changes in its antigenic structure. This was originally derived from pigs, and should technically be referred to as "pig flu".

Subtypes

There are many subtypes of avian influenza viruses, but only some strains of five subtypes have been known to infect humans: H5N1, H7N3, H7N7, H7N9, and H9N2. At least one person, an elderly woman in Jiangxi Province, China, died of pneumonia in December 2013 from the H10N8 strain, the first human fatality confirmed to be caused by that strain.

Most human cases of the avian flu are a result of either handling dead infected birds or from contact with infected fluids. It can also be spread through contaminated surfaces and droppings. While most wild birds have only a mild form of the H5N1 strain, once domesticated birds such as chickens or turkeys are infected, H5N1 can potentially become much more deadly because the birds are often in close contact. H5N1 is a large threat in Asia with infected poultry due to low hygiene conditions and close quarters. Although it is easy for humans to contract the infection from birds, human-to-human transmission is more difficult without prolonged contact. However, public health officials are concerned that strains of avian flu may mutate to become easily transmissible between humans.

Spreading of H5N1 from Asia to Europe is much more likely caused by both legal and illegal poultry trades than dispersing through wild bird migrations, being that in recent studies, there were no secondary rises in infection in Asia when wild birds migrate south again from their breeding grounds. Instead, the infection patterns followed transportation such as railroads, roads, and country borders, suggesting poultry trade as being much more likely. While there have been strains of avian flu to exist in the United States, they have been extinguished and have not been known to infect humans.

Spread of Avian Influenza

Avian influenza is most often spread by contact between infected and healthy birds, though can also be spread indirectly through contaminated equipment. The virus is found in secretions from the nostrils, mouth, and eyes of infected birds as well as their droppings. HPAI infection is spread to people often through direct contact with infected poultry, such as during slaughter or plucking. Though the virus can spread through airborne secretions, the disease itself is not an airborne disease. Highly pathogenic strains spread quickly among flocks and can destroy a flock within 28 hours; the less pathogenic strains may affect egg production but are much less deadly.

Although it is possible for humans to contract the avian influenza virus from birds, human-to-human contact is much more difficult without prolonged contact. However, public health officials are concerned that strains of avian flu may mutate to become easily transmissible between humans. Some strains of avian influenza are present in the intestinal tract of large numbers of shore birds and water birds, but these strains rarely cause human infection.

Five manmade ecosystems have contributed to modern avian influenza virus ecology: integrated indoor commercial poultry, range-raised commercial poultry, live poultry markets, backyard and hobby flocks, and bird collection and trading systems including cockfighting. Indoor commercial poultry has had the largest impact on the spread of HPAI, with the increase in HPAI outbreaks largely the result of increased commercial production since the 1990s.

Village Poultry

In the early days of the HPAI H5N1 pandemic, village poultry and their owners were frequently implicated in disease transmission. Village poultry, also known as backyard and hobby flocks, are small flocks raised under extensive conditions and often allowed free range between multiple households. However, research has shown that these flocks pose less of a threat than intensively raised commercial poultry with homogenous genetic stock and poor biosecurity. Backyard and village poultry also do not travel great distances compared to transport of intensively raised poultry and contribute less to the spread of HPAI. This initial implication of Asian poultry farmers as one broad category presented challenges to prevention recommendations as commercial strategies did not necessarily apply to backyard poultry flocks.

H5N1

The highly pathogenic influenza A virus subtype H5N1 is an emerging avian influenza virus that is causing global concern as a potential pandemic threat. It is often referred to simply as "bird flu" or "avian influenza", even though it is only one of many subtypes.

H5N1 has killed millions of poultry in a growing number of countries throughout Asia, Europe, and Africa. Health experts are concerned that the coexistence of human flu viruses and avian flu viruses (especially H5N1) will provide an opportunity for genetic material to be exchanged between species-specific viruses, possibly creating a new virulent influenza strain that is easily transmissible and lethal to humans. The mortality rate for humans with H5N1 is 60%.

Since the first human H5N1 outbreak occurred in 1997, there has been an increasing number of HPAI H5N1 bird-to-human transmissions, leading to clinically severe and fatal human infections. Because a significant species barrier exists between birds and humans, the virus does not easily spread to humans, however some cases of infection are being researched to discern whether human-to-human transmission is occurring. More research is necessary to understand the pathogenesis and epidemiology of the H5N1 virus in humans. Exposure routes and other disease transmission characteristics, such as genetic and immunological factors that may increase the likelihood of infection, are not clearly understood.

The first known transmission of H5N1 to a human occurred in Hong Kong in 1997, when there was an outbreak of 18 human cases; 6 deaths were confirmed. None of the infected people worked with poultry. After culling all of the poultry in the area, no more cases were diagnosed. In 2006, the first human-to-human transmission likely occurred when 7 members of a family in Sumatra became infected after contact with a family member who had worked with infected poultry.

Although millions of birds have become infected with the virus since its discovery, 359 people have died from H5N1 in twelve countries according to World Health Organization reports as of August 10, 2012.

As an example, the H5N1 outbreak in Thailand caused massive economic losses, especially among poultry workers. Infected birds were culled and slaughtered. The public lost confidence with the poultry products, thus decreasing the consumption of chicken products. This also elicited a ban from importing countries. There were, however, factors which aggravated the spread of the virus, including bird migration, cool temperature (increases virus survival) and several festivals at that time.

A mutation in the virus was discovered in two Guangdong patients in February 2017 which rendered it more deadly to chickens, inasmuch as it could infect every organ; the risk to humans was not increased, however.

Controversial Research

A study published in 2012, reported on research findings that allowed for the airborne transmission of H5N1 in laboratory ferrets. The study identified the 5 mutations necessary for the virus to become airborne and immediately spiked controversy over the ethical implications of making such potentially dangerous information available to the general public. The study was allowed to remain available in its entirety, though it remains a controversial topic within the scientific community.

The study in question, however, created airborne H5N1 via amino acid substitutions that largely mitigated the devastating effects of the disease. This fact was underscored by the 0% fatality rate among the ferrets infected via airborne transmission, as well as the fundamental biology underlying the substitutions. Flu viruses attach to host cells via the hemagluttinin proteins on their envelope. These hemagluttinin proteins bind to sialic acid receptors on host cells, which can fall into two categories. The sialic acid receptors can be either 2,3 or 2,6-linked, with the species of origin largely deciding receptor preference. In influenzas of avian origin 2,3-linkage is preferred, vs. influenzas of human origin in which 2,6-linkage is preferable. 2,3-linked SA receptors in humans are found predominantly in the lower respiratory tract, a fact that is the primary foundation for the deadliness of avian influenzas in humans, and also the key to their lack of airborne transmission. In the study that created an airborne avian influenza among ferrets it was necessary to switch the receptor preference of the host cells to those of 2,6-linkage, found predominantly in humans' upper respiratory tract, in order to create an infection that could shed aerosolized virus particles. Such an infection, however, must occur in the upper respiratory tract of humans, thus fundamentally undercutting the fatal trajectory of the disease.

H7N9

Influenza A virus subtype H7N9 is a novel avian influenza virus first reported to have infected humans in 2013 in China. Most of the reported cases of human infection have resulted in severe respiratory illness. In the month following the report of the first case, more than 100 people had been infected, an unusually high rate for a new infection; a fifth of those patients had died, a fifth had recovered, and the rest remained critically ill. The World Health Organization (WHO) has identified H7N9 as "an unusually dangerous virus for humans." As of June 30, 133 cases have been reported, resulting in the deaths of 43.

Research regarding background and transmission is ongoing. It has been established that many of the human cases of H7N9 appear to have a link to live bird markets. As of July 2013, there had been no evidence of sustained human-to-human transmission, however a study group headed by one of the world's leading experts on avian flu reported that several instances of human-to-human infection were suspected. It has been reported that H7N9 virus does not kill poultry, which will make surveillance much more difficult. Researchers have commented on the unusual prevalence of older males among H7N9-infected patients. While several environmental, behavioral, and biological explanations for this pattern have been proposed, as yet, the reason is unknown. Currently no vaccine exists, but the use of influenza antiviral drugs known as neuraminidase inhibitors in cases of early infection may be effective.

The number of cases detected after April fell abruptly. The decrease in the number of new human H7N9 cases may have resulted from containment measures taken by Chinese authorities, including closing live bird markets, or from a change in seasons, or possibly a combination of both factors. Studies indicate that avian influenza viruses have a seasonal pattern, thus it is thought that infections may pick up again when the weather turns cooler in China.

In the four years from early 2013 to early 2017, 916 lab-confirmed human cases of H7N9 were reported to WHO.

On 9 January 2017, the National Health and Family Planning Commission of China reported to WHO 106 cases which occurred from late November through December. 29, 2016. The cases are reported from Jiangsu (52), Zhejiang (21), Anhui (14), Guangdong (14), Shanghai (2), Fujian (2) and Hunan (1). 80 of these 106 persons have visited live poultry markets. Of these cases, there have been 35 deaths. In two of the 106 cases, human-to-human transmission could not be ruled out.

Affected prefectures in Jiangsu province closed live poultry markets in late December 2016, whereas Zhejiang, Guangdong and Anhui provinces went the route of strengthening live poultry market regulations. Travellers to affected regions are recommended to avoid poultry farms, live bird markets, and surfaces which appear to be contaminated with poultry feces. Similar sudden increases in the number of human cases of H7H9 have occurred in previous years during December and January.

Domestic Animals

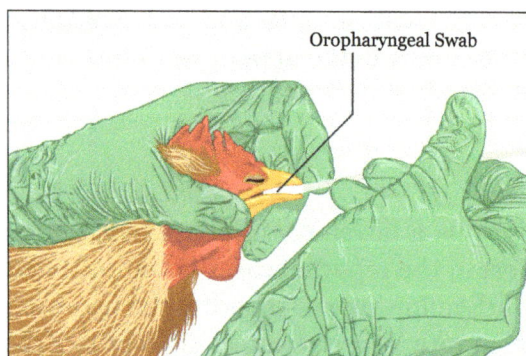

A chicken being tested for flu.

Several domestic species have been infected with and shown symptoms of H5N1 viral infection, including cats, dogs, ferrets, pigs, and birds.

Birds

Attempts are made in the United States to minimize the presence of HPAI in poultry through routine surveillance of poultry flocks in commercial poultry operations. Detection of a HPAI virus may result in immediate culling of the flock. Less pathogenic viruses are controlled by vaccination, which is done primarily in turkey flocks (ATCvet codes: QI01AA23 (WHO) for the inactivated fowl vaccine, QI01CL01 (WHO) for the inactivated turkey combination vaccine).

Seals

A recent strain of the virus is able to infect the lungs of seals.

Cats

Avian influenza in cats can show a variety of symptoms and usually lead to death. Cats are able to get infected by either consuming an infected bird or by contracting the virus from another infected cat.

Global Impact

In 2005, the formation of the International Partnership on Avian and Pandemic Influenza was announced in order to elevate the importance of avian flu, coordinate efforts, and improve disease reporting and surveillance in order to better respond to future pandemics. New networks of laboratories have emerged to detect and respond to avian flu, such as the Crisis Management Center for Animal Health, the Global Avian Influenza Network for Surveillance, OFFLU, and the Global Early Warning System for major animal diseases. After the 2003 outbreak, WHO member states have also recognized the need for more transparent and equitable sharing of vaccines and other benefits from these networks. Cooperative measures created in response to HPAI have served as a basis for programs related to other emerging and re-emerging infectious diseases.

HPAI control has also been used for political ends. In Indonesia, negotiations with global response networks were used to recentralize power and funding to the Ministry of Health. In Vietnam policymakers, with the support of the Food and Agriculture Organization of the United Nations (FAO), used HPAI control to accelerate the industrialization of livestock production for export by proposing to increase the portion of large-scale commercial farms and reducing the number of poultry keepers from 8 to 2 million by 2010.

Stigma

Backyard poultry production was viewed as "traditional Asian" agricultural practices that contrasted with modern commercial poultry production and seen as a threat to biosecurity. Backyard production appeared to hold greater risk than commercial production due to lack of biosecurity and close contact with humans, though HPAI spread in intensively raised flocks was greater due to high density rearing and genetic homogeneity. Asian culture itself was blamed as the reason why certain interventions, such as those that only looked at placed-based interventions, would fail without looking for a multifaceted solutions.

Indonesia

Press accounts of avian flu in Indonesia were seen by poultry farmers as conflating suspected cases while the public did see the accounts as informative, though many became de-sensitized to the idea of impending danger or only temporarily changed their poultry-related behavior. Rumors also circulated in Java in 2006. These tended to focus on bird flu being linked to big businesses in order to drive small farmers out of the market by exaggerating the danger of avian influenza, avian flu being introduced by foreigners to force Indonesians to purchase imported chicken and keep Indonesian chicken off the world market, and the government using avian flu as a ploy to attract funds from wealthy countries. Such rumors reflected concerns about big businesses, globalization, and a distrust of the national government in a country where "the amount of decentralization here is breathtaking" according to Steven Bjorge, a WHO epidemiologist in Jakarta in 2006.

In the context a decentralized national government that the public did not completely trust, Indonesian Health Minister Siti Fadilah Supari announced in December 2006 that her government would no longer be sharing samples of H5N1 collected from Indonesian patients. This decision came as a shock to the international community as it disrupted the Global Influenza Surveillance Network (GISN) coordinated by the WHO for managing seasonal and pandemic influenza. GISN is based on countries sharing virus specimens freely with the WHO which assesses and eventually sends these samples to pharmaceutical companies in order to produce vaccines that are sold back to these countries. Though this was initially seen as an attempt to protect national sovereignty at all costs, it was instead used for a domestic political struggle. Prior to Indonesia's dispute with the GISN, the Ministry of Health, already weak due to the decentralized nature the government, was experiencing further leakage of funding to state and non-state agencies due to global health interventions. By reasserting control over public health issues and funding by setting itself up as the sole Indonesian representative to the WHO, the Ministry of Health made itself a key player in the management of future international funds relating vaccine production and renegotiated benefits from global surveillance networks.

Economic

Approximately 20% of the protein consumed in developing countries come from poultry. In the wake of the H5N1 pandemic, millions of poultry were killed. In Vietnam alone, over 50 million domestic birds were killed due to HPAI infection and control attempts. A 2005 report by the FAO totaled economic losses in South East Asia around US $10 billion. This had the greatest impact on small scale commercial and backyard producers relative to total assets compared to industrial chains which primarily experience temporary decreases in exports and loss of consumer confidence. Some governments did provide compensation for culled poultry, it was often far below market value (close to 30% of market value in Vietnam), while others such as Cambodia provide no compensation to farmers at all.

As poultry serves as a source of food security and liquid assets, the most vulnerable populations were poor small scale farmers. The loss of birds due to HPAI and culling in Vietnam led to an average loss of 2.3 months of production and US $69–108 for households where many have an income of $2 a day or less. The loss of food security for vulnerable households can be seen in the stunting of children under 5 in Egypt. Women are another population at risk as in most regions of the world, small flocks are tended to by women. Widespread culling also resulted in the decreased enrollment of girls in school in Turkey.

Prevention

People who do not regularly come into contact with birds are not at high risk for contracting avian influenza. Those at high risk include poultry farm workers, animal control workers, wildlife biologists, and ornithologists who handle live birds. Organizations with high-risk workers should have an avian influenza response plan in place before any cases have been discovered. Biosecurity of poultry flocks is also important for prevention. Flocks should be isolated from outside birds, especially wild birds, and their waste; vehicles used around the flock should be regularly disinfected and not shared between farms; and birds from slaughter channels should not be returned to the farm.

With proper infection control and use of personal protective equipment (PPE), the chance for infection is low. Protecting the eyes, nose, mouth, and hands is important for prevention because these are the most common ways for the virus to enter the body. Appropriate personal protective equipment includes aprons or coveralls, gloves, boots or boot covers, and a head cover or hair cover. Disposable PPE is recommended. An N-95 respirator and unvented/indirectly vented safety goggles are also part of appropriate PPE. A powered air purifying respirator (PAPR) with hood or helmet and face shield is also an option.

Proper reporting of an isolated case can help to prevent spread. The Centers for Disease Control and Prevention (US) recommendation is that if a worker develops symptoms within 10 days of working with infected poultry or potentially contaminated materials, they should seek care and notify their employer, who should notify public health officials.

For future avian influenza threats, the WHO suggests a 3 phase, 5 part plan.

- Phase - Pre-pandemic:
  ∘ Reduce opportunities for human infection.
  ∘ Strengthen the early warning system.
- Phase - Emergence of a pandemic virus:
  ∘ Contain or delay spread at the source.
- Phase - Pandemic declared and spreading internationally:
  ∘ Reduce morbidity, mortality, and social disruption.
  ∘ Conduct research to guide response measures.

Vaccines for poultry have been formulated against several of the avian H5N1 influenza varieties. Control measures for HPAI encourage mass vaccinations of poultry though The World Health Organization has compiled a list of known clinical trials of pandemic influenza prototype vaccines, including those against H5N1. In some countries still at high risk for HPAI spread, there is compulsory strategic vaccination though vaccine supply shortages remain a problem.

For Village Poultry Farmers

During thr initial response to H5N1, a one size fits all recommendation was used for all poultry production systems, though measures for intensively raised birds were not necessarily appropriate for extensively raised birds. When looking at village-raised poultry, it was first assumed that the household was the unit and that flocks did not make contact with other flocks, though more effective measures came into use when the epidemiological unit was the village.

Recommendations involve restructuring commercial markets to improve biosecurity against avian influenza. Poultry production zoning is used to limit poultry farming to specific areas outside of urban environments while live poultry markets improve biosecurity by limiting the number of traders holding licenses and subjecting producers and traders to more stringent inspections. These recommendations in combination with requirements to fence and house all poultry, and to limit free ranging flocks, will eventually lead to fewer small commercial producers and backyard producers, costing livelihoods as they are unable to meet the conditions needed to participate.

A summary of reports to the World Organisation for Animal Health in 2005 and 2010 suggest that surveillance and under-reporting in developed and developing countries is still a challenge. Often, donor support can focus on HPAI control alone, while similar diseases such as Newcastle disease, acute fowl cholera, infectious laryngotracheitis, and infectious bursal disease still affect poultry populations. When HPAI tests come back negative, a lack of funded testing for differential diagnoses can leave farmers wondering what killed their birds.

Since traditional production systems require little investment and serve as a safety net for lower income households, prevention and treatment can be seen as less cost-effective than letting poultry die. Effective control not only requires prior agreements to be made with relevant government agencies, such as seen with Indonesia, they must also not unduly threaten food security.

Culling

Culling is used in order to decrease the threat of avian influenza transmission by killing potentially infected birds. The FAO manual on HPAI control recommends a zoning strategy which begins with the identification of an infected area (IA) where sick or dead birds have tested positive. All poultry in this zone are culled while the area 1 to 5 km from the outer boundary of the IA is considered the restricted area (RA) placed under strict surveillance. 2 to 10 km from the RA is the control area (CA) that serves as a buffer zone in case of spread. Culling is not recommended beyond the IA unless there is evidence of spread. The manual, however, also provides examples of how control was carried out between 2004 and 2005 to contain H5N1 where all poultry was to be stamped out in a 3 km radius beyond the infected point and beyond that a 5 km radius where all fowl was to be vaccinated. This culling method was indiscriminate as a large proportion of the poultry inside these areas were small backyard flocks which did not travel great enough distances to carry infection to adjacent villages without human effort and may have not been infected at all. Between 2004 and 2005, over 100 million chickens were culled in Asia to contain H5N1.

The risk of mass culling of birds and the resulting economic impact led farmers who were reluctant to report sick poultry. The culls often preempted actual lab testing for H5N1 as avian flu policy justified sacrificing poultry as a safeguard against HPAI spread. In response to these policies, farmers in Vietnam between 2003 and 2004 became more and more unwilling to surrender apparently healthy birds to authorities and stole poultry destined for culls as it stripped poultry of their biosocial and economic worth. By the end of 2005, the government implemented a new policy that targeted high-risk flock in the immediate vicinity of infected farms and instituted voluntary culling with compensation in the case of a local outbreak.

Not only did culling result in severe economic impacts especially for small scale farmers, culling itself may be an ineffective preventative measure. In the short-term, mass culling achieves its goals of limiting the immediate spread of HPAI, it has been found to impede the evolution of host resistance which is important for the long-term success of HPAI control. Mass culling also selects for elevated influenza virulence and results in the greater mortality of birds overall. Effective culling strategies must be selective as well as considerate of economic impacts to optimize epidemiological control and minimize economic and agricultural destruction.

People-poultry Relations

Prevention and control programs must take into account local understandings of people-poultry relations. In the past, programs that have focused on singular, place-based understandings of disease transmission have been ineffective. In the case of Northern Vietnam, health workers saw poultry as commodities with an environment that was under the control of people. Poultry existed in the context of farms, markets, slaughterhouses, and roads while humans were indirectly the primary transmitters of avian flu, placing the burden of disease control on people. However, farmers saw their free ranging poultry in an environment dominated by nonhuman forces that they could not exert control over. There were a host of nonhuman actors such as wild birds and weather patterns whose relationships with the poultry fostered the disease and absolved farmers of complete responsibility for disease control.

Attempts at singular, place-based controls sought to teach farmers to identify areas where their behavior could change without looking at poultry behaviors. Behavior recommendations by Vietnam's National Steering Committee for Avian Influenza Control and Prevention (NSCAI) were drawn from the FAO Principles of Biosecurity. These included restrictions from entering areas where poultry are kept by erecting barriers to segregate poultry from non-human contact, limits on human movement of poultry and poultry-related products ideally to transporters, and recommendations for farmers to wash hands and footwear before and after contact with poultry. Farmers, pointed to wind and environmental pollution as reasons poultry would get sick. NSCAI recommendations also would disrupt longstanding livestock production practices as gates impede sales by restricting assessment of birds by appearance and offend customers by limiting outside human contact. Instead of incorporating local knowledge into recommendations, cultural barriers were used as scapegoats for failed interventions. Prevention and control methods have been more effective when also considering the social, political, and ecological agents in play.

Marek's Disease

Leg paresis (partial paralysis) from Marek's disease.

Marek's Disease Virus (MDV) is a highly contagious viral infection that predominantly affects chickens but can also affect pheasants, quail, gamefowl and turkeys. It is one of the most common diseases that affects poultry flocks worldwide. Clinical disease is not always apparent in infected flocks, however subclinical disease is often more economically important as it reduces weight gain and egg production. Mortality rates can be very high in susceptible birds. Marek's Disease (MD) results in enlarged nerves and in tumour formation in nerve, organ, muscle and epithelial (cells that line the internal and external surfaces of the body) tissue. Clinical signs include paralysis of the legs, wings and neck, weight loss, grey iris or irregular pupil, vision impairment and the skin around feather follicles can be raised and roughened. Affected birds are immunosuppressed and as a consequence are more susceptible to other infectious diseases.

The clinical signs associated with MD can look similar to those caused by Lymphoid Leukosis and Reticuloendotheliosis, however the rareness of bursal tumours with MD helps distinguish this disease from Lymphoid Leukosis. Also, MD can develop in chickens as young as 3 weeks old, whereas Lymphoid Leukosis is typically seen in chickens older than 14 weeks old. Reticuloendotheliosis, although rare, can easily be confused with MD because both diseases feature enlarged nerves and T-cell lymphomas (a type of tumour that involves white blood cells called T-cells, which are part of the active acquired immunity system) in visceral (soft internal) organs.

Causes of Marek's Disease

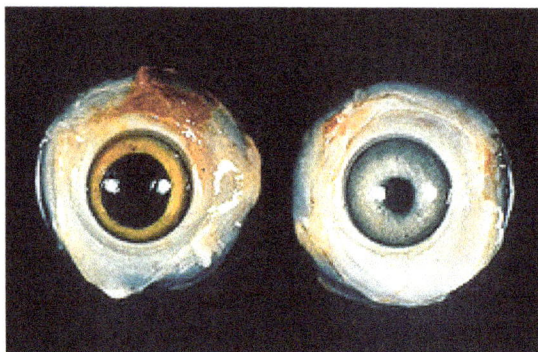

Grey iris and irregular pupil from Marek's disease.

MD is caused by a highly cell-associated (virus particles that remain attached to or within the host cell after replication) but readily transmitted herpesvirus. The route of infection is usually respiratory. The serotypes that exist are 'virulent' (disease causing) chicken isolates (serotype 1) and 'avirulent' (non-disease causing) chicken isolates (serotype 2). The avirulent virus that is commonly found in turkeys is designated serotype 3.

Serotypes are identified by reaction with serotype-specific monoclonal (clones from a single cell) antibodies or by biological characteristics such as host range, pathogenicity (severity of disease), growth rate, and plaque morphology (the physical appearance of laboratory grown viral cultures). Currently, virulent serotype 1 strains are further divided into pathotypes (classification based on the severity of disease caused by that particular strain of virus), which are often referred to as mild (m), virulent (v), very virulent (vv), and very virulent plus (vv+) MD virus strains.

The virus matures into a fully infective, enveloped virus in the cells that line the feather follicle and is released into the environment in dander (small scales from feathers which flake off and can become airborne). The virus may also be present in faeces and saliva. When cell-associated, the virus may survive for months in poultry house litter or dust and is resistant to some disinfectants.

Infected birds become carriers of the virus for life and are a source of infection to susceptible birds. MDV is less prevalent in the environment than previously thought, however, it is long lasting and remains infective in dust despite wide variations in atmospheric temperature.

Prevention and Treatment

Skin lesions from Marek's disease.

There is no treatment for MD. Vaccination is the central strategy for the prevention and control of MD. While vaccination will prevent clinical disease and reduce shedding of infective virus it will not prevent infection. Cell-associated vaccines are generally more effective than cell-free vaccines because they are neutralised less by maternal antibodies. Over time, increasingly virulent strains of MD virus have emerged, which has resulted in an ongoing need to develop new vaccines and vaccination programs to combat the disease. It was found that better protection from MD was obtained when certain combinations of serotypes were used together in a vaccine rather than one serotype alone (protective synergism). This phenomena, which is unique to MD and is strongly serotype specific, has led to the development of polyvalent vaccines (vaccines containing more than one vaccine strain). The efficacy of vaccines can be improved by strict sanitation to reduce or delay exposure and by breeding poultry for genetic resistance to MD. Vertical transmission (from parents to offspring) is not considered to be important. Vaccines administered at hatching require 1-2 weeks to produce an effective immunity, therefore exposure of chickens vaccinated at hatching to virus should be minimised during the first few days after hatching. Vaccines are also effective when administered to embryos at the 18th day of incubation. In ovo vaccination (vaccination of the embryo prior to hatching) is now performed by automated technology and is widely used for vaccination of commercial broiler chickens, mainly because of reduced labour costs and greater precision of vaccine administration.

ISSUES IN POULTRY MANAGEMENT

Roxarsone

Roxarsone is an organoarsenic compound that has been used in poultry production as a feed additive to increase weight gain and improve feed efficiency, and as a coccidiostat. As of June 2011, it was approved for chicken feed in the United States, Canada, Australia, and 12 other countries. The drug was also approved in the United States and elsewhere for use in pigs.

Its use in the United States was voluntarily ended by the manufacturers in June 2011 and has been illegal since 2013. Its use was immediately suspended in Malaysia. It was banned in Canada in August 2011. In Australia, its use in chicken feed was discontinued in 2012. Roxarsone has been banned in the European Union since 1999.

Production of Applications

Roxarsone is a derivative of phenylarsonic acid ($C_6H_5As(O)(OH)_2$). It was first reported in a 1923 British patent that described the nitration and diazotization of arsanilic acid. When blended with calcite powder, it is used in poultry feed premixes in some parts of the world. Where available, it can be purchased in 5%, 20% and 50% concentrations.

Roxarsone was marketed as 3-Nitro by Zoetis, a former subsidiary of Pfizer now a publicly traded company. In 2006, approximately one million kilograms of roxarsone were produced in the U.S.

Roxarsone has attracted attention as a source of arsenic contamination of poultry and other foods. In June 2011, the manufacturers suspended sales of roxarsone in the U.S. and Canada in response to a study by the Food and Drug Administration (FDA). The FDA found that roxarsone use was associated with elevated levels of inorganic arsenic in chicken livers. An FDA press release stated that the findings raised "concerns of a very low but completely avoidable exposure to a carcinogen."

A 2013 market basket study conducted in the United States linked the use of roxarsone and other arsenical feed additives to increased levels of inorganic arsenic in chicken breast meat, albeit at concentrations well below danger levels set in federal safety standards. Breast meat from chickens exposed to arsenical feed additives contained inorganic arsenic at the level of about two parts per billion. Organic chickens not exposed to arsenical feed additives contained about half a part per billion. Federal standards permitted concentrations of 500 parts per billion of total arsenic. The study was performed on chickens raised prior to the voluntary withdrawal of Roxarsone from the market by its manufacturer in June 2011.

Beak Trimming

Beak trimming is the removal of part of the top and bottom beak of a bird. It is also called "debeaking", although this term is inaccurate as only part of the beak is removed. It is an animal husbandry practice commonly carried out in the poultry industry. Farm managers have their flocks beak-trimmed to blunt the beaks enough to prevent the occurrence of damaging pecking. Re-trimming may also be carried out if a bird's beak grows back enough to cause pecking damage. Birds are often re-trimmed at 8–12 weeks of age to avoid this happening. Some non-trimmed adult birds may need trimming if a pecking outbreak occurs.

Beak trimmed bird.

Why is Beak Trimming done?

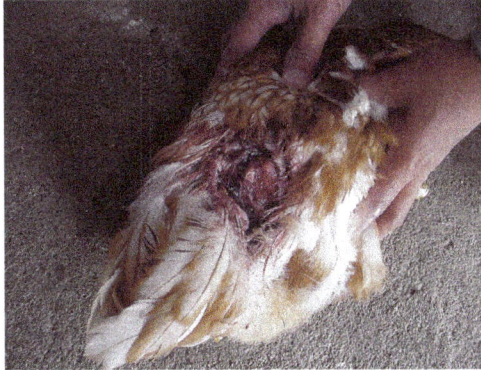

Bird seriously injured through pecking.

Beak trimming is performed early in the life of commercial hens to decrease injuries caused by cannibalism, bullying, and feather and vent pecking. Birds naturally peck at the environment and each other to investigate and work out where they fit into the flock (pecking order). This behavior can become a problem in commercial situations and many deaths have been recorded among untrimmed hens. Feather pecking and cannibalism affects all birds in all production systems. When laying birds are kept in systems that give the opportunity for aggressive birds to contact many other birds, cannibalism and feather pecking can spread rapidly through the flock and result in injuries and mortality. Mortality of up to 25–30% of the flock can occur and cause huge mortality and morbidity problems as well as financial losses to the farmer.

When is Beak Trimming done?

Infrared beak trimming machine (photo courtesy of Peter Bell).

Beak trimming is carried out at various ages depending on the preference of the farm manager. The most common ages for birds to be beak-trimmed are:

- Day-old (most common),

- 5–10 days old,

- 4–6 weeks,

- 8–12 weeks,

- Touch up trim of adult birds (mainly in alternative systems).

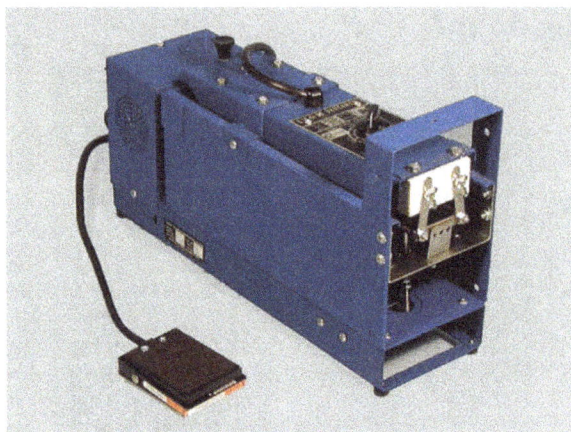
Hot blade beak trimmer.

Hot blade beak trimming is performed by contract teams, individual farmers and some large poultry companies. The majority of birds are trimmed by contract teams. Birds must be beak-trimmed by an accredited beak-trimmer to ensure that nationally agreed standards are maintained and the welfare of the birds is not compromised. The infrared treatment machine is installed by the supplier and leased by hatcheries. It is monitored and controlled by the supplier via a communication system and on-site computer.

How is Beak Trimming done?

An infrared beak trimming method, using a non-contact, high intensity, infrared energy source to treat the beak tissue, is the most common method now in use. Initially the beak surface remains intact but after a few weeks the sharp hook of the beak erodes. Experiments have also been conducted using lasers for beak trimming, however this technology is not used for beak trimming on farms. A hot blade beak trimming machine, with an electrically heated blade, is another method that has been commonly used in the past, now being surpassed by infrared.

Alternatives to Beak Trimming

Beak trimming has been banned in some European countries and others are working towards banning the practice, following an EU welfare directive on the issue. In some production, for example, 'Freedom Food Eggs' (UK), infrared beak treatment is permitted but not hot blade trimming. Even before the EU directive was released, research was being undertaken to identify practical, effective and affordable alternatives to beak trimming. Selective breeding strategies are underway to produce strains that are not cannibalistic. In addition, a number of nutritional, management and environmental strategies are being promoted as an alternative to beak trimming. The alternatives have some potential to be effective in various management situations, but there is no guarantee that cannibalism and feather pecking will be prevented.

Genetic Selection

There are large differences in feather pecking and mortality in strains indicating the potential for developing commercial strains that require less severe beak-trimming or no trimming at all. Selection for low mortality reduces propensity of birds to develop feather pecking and cannibalism.

Molecular technology has the potential for improving welfare by manipulating genes involved in the control of pecking behavior.

Light Control

Chickens have colour vision and different colours and light levels affect chicken behavior. Green and blue light improves growth and lowers age at sexual maturity, while red, orange and yellow light increases age at sexual maturity and red and orange light increase egg production. Birds are calmer in blue light. For many years it was practice to brood and rear chickens under red light to prevent cannibalism, based on the concept that red light makes it difficult for a potentially cannibalistic bird to see red blood vessels and blood on other birds. Currently, the most useful method to prevent feather pecking and cannibalism is to house birds under very dim white light. The birds cannot see each other well which is thought to reduce aggressive behavior among them. This requires light proof shedding, however low light levels can cause eye abnormalities.

Use of Devices to Restricting Vision and Beak Use

The use of spectacles (fitted to the nares of birds) controls feather pecking. It only allows birds to look to the side or down but not directly ahead. Spectacles can only be put on birds of pullet size or larger, cannot be used in cages and are easily dislodged. Red contact lenses have been used for layers as an alternative to beak-trimming. They cause eye irritation, eye infections, and abnormal behavior and are not retained well. Bitting devices have been developed for use in game birds, which are held in place by lugs inserted in the nares. The use of fitted devices as a preventative measure against feather pecking is not permitted in many countries.

Environmental Enrichment

Environmental enrichment aims to increase desirable behaviors, reduce harmful ones, sustain the birds' long-term interest, and enable them to cope with challenges. Enrichment involves increasing environmental complexity to encourage birds to interact with their environment.

Practical Enrichment Devices to Minimise Feather Pecking

A wide range of objects have been fitted to cages to enrich the environment for poultry. The 'Agrotoy' (blue plastic frame with red and blue moving parts) reduces aggression and mortality in caged layers. Likewise a small silver bell was found to attract pecking. Cereal based 'Peckablock' also reduced the amount of aggressive behavior. Adult laying hens will peck at bunches of plain white propylene string, which reduces both gentle and severe feather pecking.

Enriched Rearing Facility

Less feather pecking in layers is seen if farmers do their own rearing, provide sufficient perch space, adequate drinkers and provide high quality litter. Stimulating use of the range Infrequent and uneven use of the hen run is one of the main problems in all free range systems for laying hens. Birds do not feel safe in an open unroofed run area. When the range has cover, trees or hedges, birds are more evenly distributed and risk of feather pecking is reduced.

Use of Anti-pick Compounds

Applying anti-pick compounds (commercial anti-pick, pine tar or axle grease) to wounded areas reduces pecking. Likewise treating the everted vent of hens suffering vent trauma with a stock wound spray can prevent other birds pecking at the vent. Incidence of vent trauma can be reduced by raising flocks of birds with an even body weight. A range of predator scents and other agents are being considered for use as repellents against predators and may have application to prevent feather pecking in layers.

Using shade with free range flock.

Nutrition

The main strategy to prevent feather pecking is to provide an adequate substrate. Substrate conditions during the rearing period affect the development of feather pecking. Use of scratch grain is recommended. During the rearing period, placing semi-solid milk blocks in the house, hanging green leafy vegetables and spreading grass clippings can prevent feather pecking. An adequate amount of insoluble fibre in the layer diet is important for minimising the outbreak of cannibalism in chickens. Millrun, oat hulls, rice hulls and lucerne meal are effective sources of fibre. The physical properties of the fibre modulate the function of the gizzard making the birds calmer. Providing adequate calcium, manganese, arginine, zinc, protein, sulphur amino acids (methionine and cysteine), trytophan, B group vitamins, thiamine and dietary electrolytes minimises pecking mortality.

Beak Abrasives

Abrasive materials applied to the feed trough may enable the bird to blunt the hooked end of the beak while feeding and reduce the effectiveness of pecking. The beak blunting technique can be applied to growing pullets and during the laying period. Utilising the blunting procedure early in the rearing period may prevent the formation of the hooked end of the beak.

References

- Poultry-farming, topic: britannica.com, Retrieved 18 February, 2019

- Harris, Gardiner; Grady, Denise (9 June 2011). "Pfizer Suspends Sales of Chicken Drug". The New York Times. Retrieved 19 October 2018

- The-principles-of-poultry-husbandry, husbandry-management, production: poultryhub.org, Retrieved 19 March, 2019

- KE Nachman; PA Baron; G Raber; KA Francesconi; A Navas-Acien; DC Love (2013). "Roxarsone, Inorganic Arsenic, and Other Arsenic Species in Chicken: A U.S.-Based Market Basket Sample" (PDF). Environmental Health Perspectives. 121 (7): 818–824. Doi:10.1289/ehp.1206245. PMC 3701911. PMID 23694900

- Duck-plague, disease: northeastwildlife.org, Retrieved 20 April, 2019

- Thacker E, Janke B (February 2008). "Swine influenza virus: zoonotic potential and vaccination strategies for the control of avian and swine influenzas". The Journal of Infectious Diseases. 197 Suppl 1: S19–24. Doi:10.1086/524988. PMID 18269323

- Mareks-disease-virus-or-mdv, types-of-disease, disease: health, poultryhub.org, Retrieved 21 May, 2019

- "Infectious Bronchitis: Introduction". The Merck Veterinary Manual. 2006. Archived from the original on 22 June 2007. Retrieved 2007-06-17

Dairy Farm Management

The form of agriculture which focuses on the long-term production and/or processing of milk is known as dairy farming. Some of the technologies which are associated with dairy farming are rotolactor, bulk milk cooling tank and mixer-wagon. The diverse applications of these technologies in the management of a dairy farm have been thoroughly discussed in this chapter.

DAIRY FARM

Dairy farming is a class of agricultural, or an animal husbandry, enterprise, for long-term production of milk, usually from dairy cows but also from goats, sheep and camels, which may be either processed on-site or transported to a dairy factory for processing and eventual retail sale. Most dairy farms sell the male calves born by their cows, usually for veal production, or breeding depending on quality of the bull calf, rather than raising non-milk-producing stock. Many dairy farms also grow their own feed, typically including corn, and hay. This is fed directly to the cows, or is stored as silage for use during the winter season.

Holstein Dairy Cow.

A dairy farm is where female livestock are raised for their milk. The most common dairy animal is the cow. Dairy cattle produce a lot of milk. Farmers must collect the milk from the dairy cattle every day.

Dairy farm activities:

- Milking the cows.

- Feeding the cows.

- Helping mother cows give birth to babies (calves).

Scope and Importance

Dairying is an important source of subsidiary income to small/marginal farmers and agricultural labourers. The manure from animals provides a good source of organic matter for improving soil fertility and crop yields.

The gober gas from the dung is used as fuel for domestic purposes as also for running engines for drawing water from well. The surplus fodder and agricultural by-products are gainfully utilised for feeding the animals.

The small/marginal farmers and land less agricultural labourers play a very important role in milk production of the country. Dairy farming is now taken up as a main occupation around big urban centres where the demand for milk is high.

Scope for Dairy Farming and its National Importance

The total milk production in the country for the year 2001-02 was estimated at 84.6 million metric tonnes. At this production, the per capita availability was to be 226 grams per day against the minimum requirement of 250 grams per day.Thus, there is a tremendous scope/potential for increasing the milk production. The population of breeding cows and buffaloes in milk over 3 years of age was 62.6 million and 42.4 million, respectively (1992 census).

Management of Diary

The scheme for diary, farming should include information on land, livestock markets, availability of water, feeds, fodders, veterinary aid, breeding facilities, marketing aspects, training facilities, experience of the farmer and the type of assistance available from State Government, dairy society/union/federation.

Technical Feasibility – this would briefly include:

- Nearness of the selected area to veterinary, breeding and milk collection centre and the financing bank's branch.

- Availability of good quality animals in nearby livestock market.

- Availability of training facilities.

- Availability of good grazing ground/lands.

- Green/dry fodder, concentrate feed, medicines etc.

- Availability of veterinary aid/breeding centres and milk marketing facilities near the scheme area.

Economic Viability – this would briefly include:

- Cost of for feeds and fodders, veterinary aid, breeding of animals, insurance, labour and other overheads.

- Output costs i.e. sale price of milk, manure, gunny hags, male/female calves, other miscellaneous items etc.

Farmers

Modern and well established scientific principles, practices and skills should be used to obtain maximum economic benefits from dairy farming.

Some of the major norms and recommended practices are as follows:

- Housing:

  ◦ Construct shed on dry, properly raised ground.

  ◦ Selling of the old animals after 6-7 lactations.

- Feeding of Milch Animals:

  ◦ Feeding the animals with best feeds and fodders.

  ◦ Giving adequate green fodder in the ration.

- Milking of Animals:

  ◦ Milking the animals two to three times a day.

- Protection against Diseases:

  ◦ Be on the alert for signs of illness such as reduced feed intake, fever, abnormal discharge or unusual behavior.

- Breeding Care:

  ◦ Animal should be closely observed and keep specific record of its coming in heat, duration of heat, insemination, conception and calving.

- Care during Pregnancy:

  ◦ Give special attention to pregnant cows two months before calving by providing adequate space, feed, water etc.

- Marketing of Milk:

  ◦ Marketing milk immediately after it is drawn, keeping the time between production and marketing of the milk to the minimum.

  ◦ Production of milk produces for better storage to give more returns.

- Care of Calves:
  - Taking care of new born calf.

DAIRY CATTLE

A Holstein cow with prominent udder and
less muscle than is typical of beef breeds.

Dairy cattle (also called dairy cows) are cattle cows bred for the ability to produce large quantities of milk, from which dairy products are made. Dairy cows generally are of the species *Bos taurus*.

Historically, there was little distinction between dairy cattle and beef cattle, with the same stock often being used for both meat and milk production. Today, the bovine industry is more specialized and most dairy cattle have been bred to produce large volumes of milk.

Management

Cows on a dairy farm in Maryland.

Dairy cows may be found either in herds or dairy farms where dairy farmers own, manage, care for, and collect milk from them, or on commercial farms. Herd sizes vary around the world depending on landholding culture and social structure. The United States has an estimated 9 million cows in around 75,000 dairy herds, with an average herd size of 120 cows. The number of small herds is falling rapidly with the 3,100 herds with over 500 cows producing 51% of U.S. milk in 2007. The

United Kingdom dairy herd overall has nearly 1.5 million cows, with about 100 head reported on an average farm. In New Zealand, the average herd has more than 375 cows, while in Australia, there are approximately 220 cows in the average herd.

The United States dairy herd produced 84.2 billion kilograms (185.7 billion pounds) of milk in 2007, up from 52.9 billion kilograms (116.6 billion pounds) in 1950, yet there were only about 9 million cows on U.S. dairy farms—about 13 million fewer than there were in 1950. The top breed of dairy cow within Canada's national herd category is Holstein, taking up 93% of the dairy cow population, have an annual production rate of 10,257 kilograms (22,613 pounds) of milk per cow that contains 3.9% butter fat and 3.2% protein.

Dairy farming, like many other livestock raring, can be split into intensive and extensive management systems.

Intensive systems focus towards maximum production per cow in the herd. This involve formulating their diet to provide ideal nutrition and housing the cows in a confinement system such as free stall or tie stall. These cows are housed indoors throughout their lactation and may be put to pasture during their 60-day dry period before ideally calving again. Free stall style barns involve cattle loosely housed where they can have free access to feed, water, and stalls but are moved to another part of the barn to be milked multiple times a day. In a tie stall system, the milking units are brought to the cows during each milking. These cattle are tethered within their stalls with free access to water and feed are provided. In extensive systems, cattle are mainly outside on pasture for most of their lives. These cattle are generally lower in milk production and are herded multiple times daily to be milked. The systems used greatly depends on the climate and available land of the region of which the farm is situated.

To maintain lactation, a dairy cow must be bred and produce calves. Depending on market conditions, the cow may be bred with a "dairy bull" or a "beef bull." Female calves (heifers) with dairy breeding may be kept as replacement cows for the dairy herd. If a replacement cow turns out to be a substandard producer of milk, she then goes to market and can be slaughtered for beef. Male calves can either be used later as a breeding bull or sold and used for veal or beef. Dairy farmers usually begin breeding or artificially inseminating heifers around 13 months of age. A cow's gestation period is approximately nine months. Newborn calves are separated from their mothers quickly, usually within three days, as the mother/calf bond intensifies over time and delayed separation can cause extreme stress on both cow and calf.

Domestic cows can live to 20 years; however, those raised for dairy rarely live that long, as the average cow is removed from the dairy herd around age six and marketed for beef. In 2014, approximately 9.5% of the cattle slaughtered in the U.S. were culled dairy cows: cows that can no longer be seen as an economic asset to the dairy farm. These animals may be sold due to reproductive problems or common diseases of milk cows such as mastitis and lameness.

Calf

Most heifers (female calves) are kept on farm to be raised as a replacement heifer, a female that is bred and enters the production cycle. Market calves are generally sold at two weeks of age and bull calves may fetch a premium over heifers due to their size, either current or potential. Calves may be sold for veal, or for one of several types of beef production, depending on available local crops and markets.

Such bull calves may be castrated if turnout onto pastures is envisaged, to make them less aggressive. Purebred bulls from elite cows may be put into progeny testing schemes to find out whether they might become superior sires for breeding. Such animals can become extremely valuable.

Most dairy farms separate calves from their mothers within a day of birth to reduce transmission of disease and simplify management of milking cows. Studies have been done allowing calves to remain with their mothers for 1, 4, 7 or 14 days after birth. Cows whose calves were removed longer than one day after birth showed increased searching, sniffing and vocalizations. However, calves allowed to remain with their mothers for longer periods showed weight gains at three times the rate of early removals as well as more searching behavior and better social relationships with other calves.

After separation, some young dairy calves subsist on commercial milk replacer, a feed based on dried milk powder. Milk replacer is an economical alternative to feeding whole milk because it is cheaper, can be bought at varying fat and protein percentages, and is typically less contaminated than whole milk when handled properly. Some farms pasteurize and feed calves milk from the cows in the herd instead of using replacer. A day-old calf consumes around 5 liters of milk per day.

Cattle are social animals; their ancestors tended to live in matriarchal groups of mothers and off-spring. The formation of "friendships" between two cows is common and long lasting. Traditionally individual housing systems were used in calf rearing, to reduce the risk of disease spread and provide specific care. However, due to their social behavior the grouping of offspring may be better for the calves' overall welfare. Social interaction between the calves can have a positive effect on their growth. It has been seen that calves housed in grouped penning were found to eat more feed than those in single pens, suggesting social facilitation of feeding behavior in the calves. Play behavior in pre-weaned dairy calves has also been suggested to help build social skills for later in life. It has been seen that those reared in grouped housing are more likely to become the dominant cattle in a new combination of animals. These dominant animals have a priority choice of feed or lying areas and are generally stronger animals. Due to these reasons, it has become common practice to group or pair calves in their housing. It has become common within Canada to see paired or grouped housing in outdoor hutches or within an indoor pack penning.

Bull

A bull calf with high genetic potential may be reared for breeding purposes. It may be kept by a dairy farm as a herd bull, to provide natural breeding for the herd of cows. A bull may service up to 50 or 60 cows during a breeding season. Any more and the sperm count declines, leading to cows "returning to service" (to be bred again). A herd bull may only stay for one season, as when most bulls reach over two years old their temperament becomes too unpredictable.

Bull calves intended for breeding commonly are bred on specialized dairy breeding farms, not production farms. These farms are the major source of stocks for artificial insemination.

Milk Production Levels

The dairy cow produces large amounts of milk in its lifetime. Production levels peak at around 40 to 60 days after calving. Production declines steadily afterwards until milking is stopped at about 10 months. The cow is "dried off" for about sixty days before calving again. Within a 12 to 14-month inter-calving cycle, the milking period is about 305 days or 10 months long. Among many variables, certain breeds produce more milk than others within a range of around 6,800 to 17,000 kg (15,000 to 37,500 lb) of milk per year.

Dairy cattle in Mangskog, Sweden.

The Holstein Friesian is the main breed of dairy cattle in Australia, and said to have the "world's highest" productivity, at 10000 L of milk per year. The average for a single dairy cow in the US in 2007 was 9,164 kg (20,204 lb) per year, excluding milk consumed by her calves, whereas the same average value for a single cow in Israel was reported in the Philippine press to be 12,240 kg in 2009. High production cows are more difficult to breed at a two-year interval. Many farms take the view that 24 or even 36 month cycles are more appropriate for this type of cow.

Dairy Cows, Collins Center, New York.

Dairy cows may continue to be economically productive for many lactation cycles. In theory a longevity of 10 lactations is possible. The chances of problems arising which may lead to a cow being culled are high, however; the average herd life of US Holstein is today fewer than 3 lactations. This requires more herd replacements to be reared or purchased. Over 90% of all cows are slaughtered for 4 main reasons:

- Infertility: Failure to conceive and reduced milk production.

  Cows are at their most fertile between 60 and 80 days after calving. Cows remaining "open" (not with calf) after this period become increasingly difficult to breed, which may be due to poor health. Failure to expel the afterbirth from a previous pregnancy, luteal cysts, or metritis, an infection of the uterus, are common causes of infertility.

- Mastitis: A persistent and potentially fatal mammary gland infection, leading to high somatic cell counts and loss of production.

  Mastitis is recognized by a reddening and swelling of the infected quarter of the udder and the presence of whitish clots or pus in the milk. Treatment is possible with long-acting antibiotics but milk from such cows is not marketable until drug residues have left the cow's system, also called withdrawal period.

- Lameness: Persistent foot infection or leg problems causing infertility and loss of production.

  High feed levels of highly digestible carbohydrate cause acidic conditions in the cow's rumen. This leads to laminitis and subsequent lameness, leaving the cow vulnerable to other foot infections and problems which may be exacerbated by standing in faeces or water soaked areas.

- Production: Some animals fail to produce economic levels of milk to justify their feed costs. Production below 12 to 15 L of milk per day is not economically viable.

Cow longevity is strongly correlated with production levels. Lower production cows live longer than high production cows, but may be less profitable. Cows no longer wanted for milk production are sent to slaughter. Their meat is of relatively low value and is generally used for processed meat. Another factor affecting milk production is the stress the cow is faced with. Psychologists at the University of Leicester, UK, analyzed the musical preference of milk cows and found out that music actually influences the dairy cow's lactation. Calming music can improve milk yield, probably because it reduces stress and relaxes the cows in much the same way as it relaxes humans.

Cow Comfort and its Effects on Milk Production

Certain behaviors such as eating, rumination, and lying down can be related to the health of the cow and cow comfort. These behaviors can also be related to the productivity of the cows. Likewise, stress, disease, and discomfort negatively effect milk productivity. Therefore, it can be said that it is in the best interest of the farmer to increase eating, rumination, and lying down and decrease stress, disease, and discomfort to achieve the maximum productivity possible. Also, estrous behaviors such as mounting can be a sign of cow comfort, since if a cow is lame, nutritionally deficient, or housed in an over crowded barn, its estrous behaviors is altered.

Feeding behaviors are important for the dairy cow, as feeding is how the cow ingests dry matter. However, the cow must ruminate to fully digest the feed and utilize the nutrients in the feed. Dairy cows with good rumen health are likely to be more profitable than cows with poor rumen health—as a healthy rumen aids in digestion of nutrients. An increase in the time a cow spends ruminating is associated with the increase in health and an increase in milk production. The productivity of dairy cattle is most efficient when the cattle have a full rumen. Also, the standing action while feeding after milking has been suggested to enhance udder health. The delivery of fresh feed while the cattle are away for milking stimulates the cattle to fed upon return, potentially reducing the prevalence of mastitis as the sphincters have time to close while standing This makes the pattern of feeding directly after being milked an ideal method of increasing the efficiency of the herd.

Cows have a high motivation to lie down so farmers should be conscious of this, not only because they have a high motivation to lie down, but also because lying down can increase milk yield. When the lactating dairy cow lies down, blood flow is increased to the mammary gland which in return results in a higher milk yield.

To ensure that the dairy cows lie down as much as needed, the stalls must be comfortable. Put very simply, a stall should have a rubber mat, bedding, and be large enough for the cow to lie down and get up comfortably. Signs that the stalls may not be comfortable enough for the cows are the cows are standing, either ruminating or not, instead of lying down, or perching, which is when the cow has its front end in the stall and their back end out of the stall.

There are two types of housing systems in dairy production, free style housing and tie stall. Free style housing is where the cow is free to walk around and interact with its environment and other members of the herd. Tie stall housing is when the cow is chained to a stantion stall with the milking units and feed coming to them.

By-products and Processing

Pasteurization is the process of heating milk to a high enough temperature for a short period of time to kill the microbes in the milk and increase keep time and decrease spoilage time. By killing the microbes, decreasing the transmission of infection, and elimination of enzymes the quality of the milk and the shelf life increases. Pasteurization is either completed at 63 °C for thirty minutes or a flash pasteurization is completed for 15 seconds at 72 °C. By-products of milk include butterfat, cream, curds, and whey. Butterfat is the main lipid in milk. The cream contains 18–40% butterfat. The industry can be divided into 2 market territories; fluid milk and industrialized milk such as yogurt, cheeses, and ice cream.

Whey protein makes up about 20% of milk's protein composition and is separated for the casein (80% of milk's protein make up) during the process of curdling cheese. This protein is commonly used in protein bars, beverages and concentrated powder, due to its high quality amino acid profile. It contains levels of both essential amino acids as well as branched that are above those of soy, meat, and wheat. "Diafiltered" milk is a process of ultrafiltration of the fluid milk to separate lactose and water from the casein and whey proteins. This process allows for more efficiency in cheese making and gives the potential to produce low-carb dairy products.

Reproduction

Since the 1950s, artificial insemination (AI) is used at most dairy farms; these farms may keep no bull. Artificial insemination uses estrus synchronization to indicate when the cow is going through ovulation and is susceptible to fertilization. Advantages of using AI include its low cost and ease compared to maintaining a bull, ability to select from a large number of bulls, elimination of diseases in the dairy industry, improved genetics and improved animal welfare. Rather than a large bull jumping on a smaller heifer or weaker cow, AI allows the farmer to complete the breeding procedure within 5 minutes with minimum stress placed on the individual female's body.

Dairy cattle are polyestrous, meaning they cycle continuously throughout the year. They tend to be on a 21 day estrus cycle. However for management purposes, some operations use synthetic hormones to synchronize their cows or heifers to have them breed and calve at the ideal times.

These hormones are short term and only used when necessary. For example, one common proto-col for synchronization involves an injection of GnRH (gonadotrophin releasing hormone). which increases the levels of follicle stimulating hormone and luteinizing hormone in the body. Then, seven days later prostaglandin F2-alpha is injected, followed by another GnRH injection 48 hours later. This protocol causes the animal to ovulate 24 hours later.

Estrus is often called standing heat in cattle and refers to the time in their cycle where the female is receptive towards the male. Estrus behavior can be detected by an experienced stockman. These behaviors can include standing to be mounted, mounting other cows, restlessness, decreased milk production, and decreased feed intake.

More recently, embryo transfer has been used to enable the multiplication of progeny from elite cows. Such cows are given hormone treatments to produce multiple embryos. These are then 'flushed' from the cow's uterus. 7–12 embryos are consequently removed from these donor cows and transferred into other cows who serve as surrogate mothers. This results in between three and six calves instead of the normal single or (rarely) twins.

Hormone Use

Farmers in some countries sometimes administer hormone treatments to dairy cows to increase milk production and reproduction.

About 17% of dairy cows in the United States are injected with Bovine somatotropin, also called recombinant bovine somatotropin (rBST), recombinant bovine growth hormone (rBGH), or arti-ficial growth hormone. The use of this hormone increases milk production by 11%–25%. The U.S. Food and Drug Administration (FDA) has ruled that rBST is harmless to people. The use of rBST is banned in Canada, parts of the European Union, as well as Australia and New Zealand.

In Canada, Canadian Dairy farmers have high screening procedures they have to go through every time the milk is retrieved from the farm; if the regulations are not met the milk does not get loaded onto the truck for further processing. There is to be no medication or hormones in the milk for safety reasons.

Nutrition

Dairy cattle at feeding time.

Nutrition plays an important role in keeping cattle healthy and strong. Implementing an adequate nutrition program can also improve milk production and reproductive performance. Nutrient re-quirements may not be the same depending on the animal's age and stage of production.

Forages, which refer especially to hay or straw, are the most common type of feed used. Cereal grains, as the main contributors of starch to diets, are important in meeting the energy needs of dairy cattle. Barley is an excellent source of balanced amounts of protein, energy, and fiber.

Ensuring adequate body fat reserves is essential for cattle to produce milk and also to keep reproductive efficiency. However, if cattle get excessively fat or too thin, they run the risk of developing metabolic problems and may have problems with calving. Scientists have found that a variety of fat supplements can benefit conception rates of lactating dairy cows. Some of these different fats include oleic acids, found in canola oil, animal tallow, and yellow grease; palmitic acid found in granular fats and dry fats; and linolenic acids which are found in cottonseed, safflower, sunflower, and soybean.

Using by-products is one way of reducing the normally high feed costs. However, lack of knowledge of their nutritional and economic value limits their use. Although the reduction of costs may be significant, they have to be used carefully because animal may have negative reactions to radical changes in feeds, (e.g. fog fever). Such a change must then be made slowly and with the proper follow up.

Breeds

According to the Purebred Dairy Cattle Association, PDCA, there are 7 major dairy breeds in the United States. These are: Holstein, Brown Swiss, Guernsey, Ayrshire, Jersey, Red and White, and Milking Shorthorn.

Holstein cows either have distinct white and black markings, or distinct red and white markings. Holstein cows are the biggest of all U.S. dairy breeds. A full mature Holstein cow usually weighs around 700 kilograms (1,500 lb) and is 147 centimetres (58 in) tall at the shoulder. They are known for their outstanding milk production among the main breeds of dairy cattle. An average Holstein cow produces around 10,000 kilograms (23,000 lb) of milk each lactation. Of the 9 million dairy cows in the U.S., approximately 90% of them are of the Holstein descent. The top breed of dairy cow within Canada's national herd category is Holstein, taking up 93% of the dairy cow population, have a production rate of 10,257 kilograms (22,613 lb) of milk per cow that contains 3.9% butter fat and 3.2% protein.

Brown Swiss cows are widely accepted as the oldest dairy cattle breed, originally coming from a part of northeastern Switzerland. Some experts think that the modern Brown Swiss skeleton is similar to one found that looks to be from around the year 4000 BC Also, there is evidence that monks started breeding these cows about 1000 years ago.

The Ayrshire breed first originated in the County of Ayr in Scotland. It became regarded as a well established breed in 1812. The different breeds that were crossed to form the Ayrshire are not exactly known. However, there is evidence that several breeds were crossed with the native cattle to create the breed.

Guernsey cows originated just off the coast of France on the small Isle of Guernsey. The breed was first known as a separate breed around 1700. Guernseys are known for their ability to produce very high quality milk from grass. Also, the term "Golden Guernsey" is very common as Guernsey cattle produce rich, yellow milk rather than the standard white milk other cow breeds produce.

The Jersey breed of dairy cow originated on a small island located off the coast of France called Jersey. Despite being one of the oldest breeds of dairy cattle they now only occupy 4% of the Canadian National Herd. Purebred Jersey cows, according to available data, have been in the UK area since about the year 1741. When they were first bred in this area, they were not known as Jerseys, but rather as the related Alderneys. The period between 1860 and around 1914 was a popular time for Jerseys. In this time span, many countries other than the United States started importing this breed, including Canada, South Africa, and New Zealand, among others.

Among the smallest of the dairy breeds, the average Jersey cow matures at approximately 410 kilograms (900 lb), with a typical weight range between 360 and 540 kilograms (800–1,200 lb). According to North Dakota State University, the fat content of the Jersey cow's milk is 4.9 percent. It is also the highest in protein, at 3.8 percent. This high fat content means the milk is often used for making ice cream and cheeses. According to the American Jersey Cattle Association, Jerseys are found on 20 percent of all US dairy farms and are the primary breed in about 4 percent of dairies.

Amongst the *Bos indicus*, the most popular dairy breed in the world is Sahiwal of the Indian subcontinent. It does not give as much milk as the Taurine breeds, but it is by far the most suitable breed for warmer climates. Australian Friesian Sahiwal and Australian Milking Zebu have been developed in Australia using Sahiwal genetics. Gir, another of the *Bos indicus* breeds, has been improved in Brazil for its milk production and is widely used there for dairy.

COW LACTATION CYCLES

Poor feeding management of cows can lead to shorter, lower yielding lactations and increase calving interval.

Lactation Cycle

Cows must calve to produce milk and the lactation cycle is the period between one calving and the next.

The cycle is split into four phases, the early, mid and late lactation (each of about 120 days, or d) and the dry period (which should last as long as 65 d). In an ideal world, cows calve every 12 months. A number of changes occur in cows as they progress through different stages of lactation.

As well as variations in milk production, there are changes in feed intake and body condition, and stage of pregnancy. Figure presents the interrelationships between feed intake, milk yield and live weight for a Friesian cow with a 14 month inter-calving interval, hence a 360 d lactation.

Following calving, a cow may start producing 10 kg/d of milk, rise to a peak of 20 kg/d by about 7 weeks into lactation then gradually fall to 5 kg/d by the end of lactation.

Although her maintenance requirements will not vary, she will need more dietary energy and protein as milk production increases then less when production declines. However to regain body condition in late lactation, she will require additional energy.

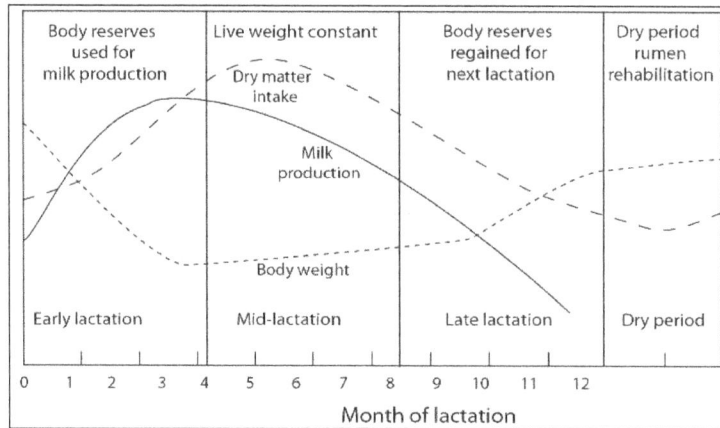

Dry matter intake, milk yield and live weight
changes in a cow during her location cycle.

Cows usually use their own body condition for about 12 weeks after calving, to provide energy in addition to that consumed. The energy released is used to produce milk, allowing them to achieve higher peak production than would be possible from their diet alone.

To do this, cows must have sufficient body condition available to lose, and therefore they must have put it on late in the previous lactation or during the dry period.

From Calving to Peak Lactation

Milk yield at the peak of lactation sets up the potential milk production for the year; one extra kg per day at the peak can produce an extra 200 kg/cow over the entire lactation.

There are a number of obstacles to feeding the herd well in early lactation to maximise the peak. The foremost of these is voluntary food intake.

At calving, appetite is only about 50 to 70 per cent of the maximum at peak intake. This is because during the dry period, the growing calf takes up space, reducing rumen volume and the density and size of rumen papillae is reduced.

After calving, it takes time for the rumen to "stretch" and the papillae to regrow. It is not until weeks 10-12 that appetite reaches its full potential.

Peak Lactation to Peak Intake

Following peak lactation, cows' appetites gradually increase until they can consume all the nutrients required for production, provided the diet is of high quality. From figure, cows tend to maintain weight during this stage of their lactation.

Mid and Late Lactation

Although energy required for milk production is less demanding during this period because milk production is declining, energy is still important because of pregnancy and the need to build up body condition as an energy reserve for the next lactation. It is generally more efficient to improve the condition of the herd in late lactation rather than in the dry period.

Dry Period

Maintaining (or increasing) body condition during the dry period is the key to ensuring cows have adequate body reserves for early lactation.

If cows calve with adequate body reserves, they can cycle within two or three months after calving. If cows calve in poor condition, milk production suffers in early lactation because body reserves are not available to contribute energy.

In fact, dietary energy can be channelled towards weight gain rather being made available from the desired weight loss. For this reason, high feeding levels in early lactation cannot make up for poor body condition at calving.

Persistency of Milk Production throughout Lactation

The two major factors determining total lactation yield are peak lactation and the rate of decline from this peak. In temperate dairy systems, total milk yield for 300 day lactation can be estimated by multiplying peak yield by 200.

Hence a cow peaking at 20 litres per day (L/d) should produce 4000 L/lactation, while a peak of 30 L/d equates to a 6000 L full lactation milk yield. This is based on a rate of decline of 7 to 8 per cent per month from peak yield, that is every month the cow produces, on average, 7 to 8 per cent of peak yield less than in the previous month.

This level of persistency is the target for well managed, pasture-based herds in temperate regions.

Actual values can vary from 3 to 4 per cent per month in fully fed, lot fed cows to 12 per cent or more per month in very poorly fed cows, for example during a severe dry season following a good wet season in the tropics.

The rate of decline from peak, or persistency, depends on:

* Peak milk yield.
* Nutrient intake following peak yield.
* Body condition at calving.
* Other factors such as disease status and climatic stress.

Generally speaking, the higher the milk yield at peak, the lower its persistency in percentage terms. Underfeeding of cows immediately post-calving reduces peak yield but also has adverse effects on persistency and fertility. Dairy cows have been bred to utilise body reserves for additional milk production, but high rates of live weight loss will delay the onset of oestrus.

Underfeeding of high genetic merit cows in early lactation is one of the biggest nutritionally in-duced problems facing many small holder farmers in the humid tropics, because they often do not have the necessary improvements in feeding systems to utilise high genetic potential.

If imported high genetic quality cows are not well fed, milk production is compromised, but of more importance, they will not cycle until many months post-calving.

Theoretical Models of Lactation Persistency

Table and figure below present data for milk yield over 300 day lactations in cows with various peak milk yields and lactation persistencies.

Such data provides the basis of herd management guidelines for dairy systems with 12 month calving intervals. Depending on herd fertility, hence target lactation lengths, similar guidelines could be developed for 15 or 18 month calving intervals.

Table: Effect of peak milk yield and persistency on 300 day location yields.

Peak yield (L/d)	Persistency (% mth)	Monthly milk decline (L/d)	Full lact yield (L)	Average milk yield (L/d)
	8	1.2	2980	9.9
15	10	1.5	2650	8.9
	12	1.8	2330	7.8
	8	1.6	3970	13.2
20	10	2.0	3540	11.8
	12	2.4	3110	10.4
	8	2.0	4960	16.6
25	10	1.5	4420	14.8
	12	3.0	3885	13.0

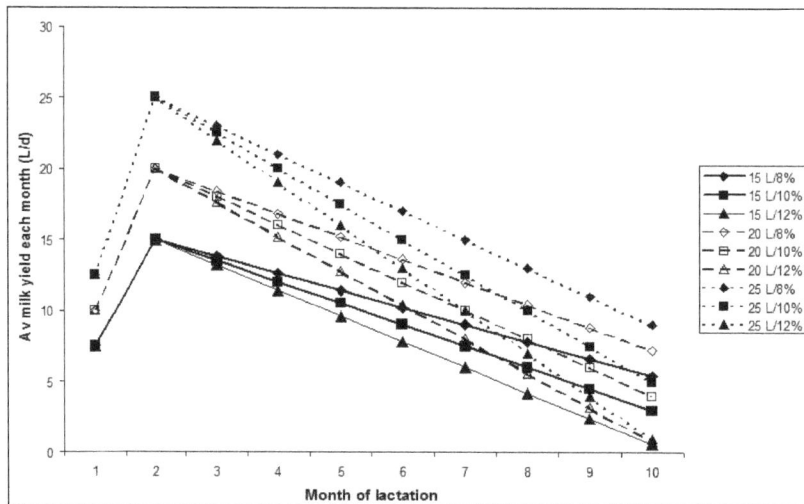

Milk Yields each month for cows varying in peak yield and persistency.
Legend shows peak yield (L/d) and persistency (% decline/mth).

Small holder dairy farms in the humid tropics with good feeding and herd management should be able to achieve 15 L/day peak yield, and for those with high genetic merit cows, 20 or 25 L/day is realistic.

Lactation persistencies of less than 8 per cent per month may be achievable in tropical dairy feedlots but more realistic persistencies are the 8 to 12 per cent per month presented in the table and figure above.

Virtually every small holder farmer records daily milk yield of his or her cows, so they know peak yield and can easily determine the monthly rate of decline, providing a simple monitoring tool to assess their level of feeding management.

Unless feeding management can be improved, it may be better in the long run to import cows of lower genetic merit.

For example, importers may request "5000 L cows" (that is cows that peak at 25 L/day under good feeding management, with a persistency of 8 per cent/mth).

If, through poor feeding, their persistency is reduced to 12 per cent per month, 300 d lactation yields are only 3900 L and they do not cycle for many months after calving, "4000 L cows" may be a better investment. From table, such cows would produce similar milk yields if they could be fed to 8 per cent per month milk persistency and they are more likely to cycle earlier.

Impacts of Short Lactation Length

Poor feeding management of potentially high yielding cows can create many problems. Lactation anoestrus can occur as the cows are forced to utilise more of their body reserves in early lactation. This can lead to low peak milk yields and shortened lactation lengths.

Cows will dry off prematurely if they receive insufficient feed nutrients to maintain viable processes of milk production in their mammary tissue.

The impact of decreasing lactation lengths on 300 day lactation milk yields and average daily milk yields are presented in table. These data are based on the same persistency data used in table. The penalties for these shortened lactation lengths are presented in table.

Compared to 10 month lactations, inherently poor yielding cows with low peak milk yields can lose 20 to 160 L milk through only 9 months milking or 90 to 360 L milk if only milking for 8 months.

Following higher peak milk yields, this will increase to penalties of 30 to 270 L milk for 9 month to 120 to 600 L for 8 month lactation lengths. This can have a big effect on the herd's rolling herd average which can be reduced by 0.3 to 2.0 L/cow/day for the extreme values presented in table below.

Table: Effect of peak milk yield, persistency and lactation length on 300 d lactation yields.

Lact length		300days		270 Days		240 Days	
Peak yield (L/d)	Persistency (%mth)	300 d Lact yield(L)	Av milk Yield (L/d)	300 d Lact yield (L)	Av milk Yield (L/d)	300 d Lact yield(L)	Av milk Yield (L/d)
	8	2980	9.9	2820	9.4	2620	8.7
15	10	2650	8.9	2560	8.6	2430	8.1
	12	2330	7.8	2313	7.7	2240	7.5
	8	3970	13.2	3760	12.5	3490	11.6
20	10	3540	11.8	3420	11.4	3240	10.8
	12	3110	10.4	3084	10.3	2990	10.0

	8	4960	16.6	4690	15.7	4360	14.6
25	10	4420	14.0	4280	14.3	4050	13.5
	12	3885	13.0	3860	12.9	3740	12.5

Table: Penalties for shortened lactation length, compared to 300 days.

Lact length		270 Days		240 Days	
Peak yield (L/d)	Persistency (%mth)	300 d Lact yield(L)	Av milk Yield (L/d)	300 d Lact yield(L)	Av milk Yield (L/d)
15	8	162	0.5	360	1.2
	10	90	0.3	225	0.8
	12	18	0.1	90	0.3
20	8	216	0.4	480	1.6
	10	120	0.7	300	1.0
	12	24	0.1	120	0.4
25	8	270	0.9	600	2.0
	10	150	0.5	375	1.3
	12	30	0.1	150	0.5

These tables are based on 300 day lactation lengths, that is under an ideal situation where cows calve down every 12 months.

Inter-calving intervals are more likely to be 13, 14 or 15 months, hence lactation lengths should be even longer than 300 days.

Ideally cows should be managed to have a two month dry period to allow the mammary tissue to recuperate before the next lactation. However, lactation lengths of just 8 months followed by dry periods of another 8 months are all too common in many tropical small holder dairy farms. This then equates to only 50 per cent of the adult cows milking at any one time.

DAIRY FARMING TECHNOLOGIES

One trend within the dairy industry today is the increased incorporation of computers into the day-to-day management of dairy herds. Computer use aids producers in the management of their herds and allows for better record keeping. The next step in on-farm technology use is the use of precision dairy farming technologies. Precision dairy farming technologies are instruments to measure physiological, behavioral, and production indicators for individual animals. In the past few decades, precision dairy technologies have become available, which when implemented properly, make dairy producers more efficient. Efficient dairy farmers have a better chance at staying competitive and financially secure as milk prices become increasingly volatile. When financially sustainable technologies and good management practices are used in unison, new levels of farm productivity become obtainable.

Many different kinds of technologies exist in the market today. The sheer number of various technologies and the information produced by them can be overwhelming and confusing. Information about these technologies exists but can often times be hard to find or difficult to understand.

Unfamiliarity with technologies and how they work can become an obstacle to overcome, especially when comparing and contrasting technologies. This fact-sheet will list several of these technologies and how they accomplish their goals.

The decision to adopt a technology depends on factors such as management style, familiarity with computers, ease of use, type of housing system, and perceived benefit to cost ratio. Technologies monitoring various parameters are available to farmers and often these technologies fall into categories including:

- Nutrition- individual feeders, mixing equipment, and water supplies.

- Production- in-parlor controls and monitors.

- Animal health- mastitis, rumen health, metabolic disorders, and body temperature.

- Fertility- estrus and calving detection.

- Environmental- temperature and milk line vacuum.

Additional categories exist and as technology continues to advance, new kinds of technologies monitoring new parameters will come into existence.

This list includes information pertaining to each technology and outlines its location relative to the cow, what it measures, and reports generated from those measurements. Below is an explanation of the location and parameters measured by each technology.

Parameter Measured

- Animal position/Location - locates animals within a facility or pasture.

- Blood content in milk - determines general udder health and SCC through monitoring of blood content in milk.

- Body weight - detects changes in animals' total weight to evaluate the health and efficiency of feed energy conversion to milk.

- Cow activity – collects two-dimensional and three-dimensional animal movement data to monitor health and estrus behavior.

- Feeding behavior- monitors eating events and durations to observe health, behavior, or estrus.

- Fertility hormones (e.g. Progesterone) - determines reproductive performance by quantifying the amounts of circulating reproductive hormones and determining estrus or pregnancy.

- Jaw movement/Chewing activity - tracks eating events and rumination as indicators of animal health, nutritional, and reproductive performance.

- Lameness- monitors animal mobility.

- LDH (Lactate dehydrogenase) - detects changes in level of LDH to predict udder health.

- Lying/Standing behavior- collects level of activity and three-dimensional movement of animals for monitoring animal health, comfort, and activity.

- Mastitis- detects changes in whole udder or individual quarters for detecting and quantifying level of infection.

- Milk components (e.g. fat, protein) - quantifies and monitors milk constituents.

- Milk conductivity- determines changes in milk ionic concentrations to indicate milk fat, protein, lactose, or somatic cell count quantities.

- Milk flow- milk produced in a given amount of time that is generally used for automatic take-offs, preserving teat ends.

- Milk time- monitors the time a cow or each individual quarter spends milking.

- Milk yield- amount of milk given by individuals or groups of animals.

- Rumen pH- health through monitoring of rumen pH.

- Rumination- determines health and estrus by quantifying eructation.

- Somatic cell count- determines udder health through the somatic cell count of milk originating from all or individual quarters of the cow.

- Standing heat- records events where a cow or heifer stands to be mounted.

- Temperature- determines estrus, calving, stress, or general health of an animal by monitoring temperature in the ear, rumen, vagina, or milk.

- Vacuum in milk line- preserves teat ends by monitoring fluctuations in vacuum strength present at the teat ends.

Rotolactor

The Rotolactor is the first invention for milking a large number of cows successively and largely automatically, using a rotating platform. It was developed by the Borden Company in 1930, and is known today in the dairy industry as the "rotary milking parlor".

The Rotolactor was the first invention for milking a large number of cows using a rotating platform. It was invented by Henry W. Jeffers. The Rotolactor was initially installed in a "lactorium," a building specifically designed for milking cows, in Plainsboro, New Jersey. The rotating mechanical milking machine was first used by the Walker-Gordon Laboratories dairy and was put into operation on November 13, 1930.

Jeffers conceived the idea for the Rotolactor in 1913 as a cost-cutting and labor-saving method for milking a large number of cows. Development of the project was put on hold during World War I. In 1928, the Walker-Gordon Laboratories dairy was purchased by the Borden Company, and Rotolactor development resumed in earnest. Borden provided $200,000 in 1929 for building the Rotolactor at the Walker-Gordon Laboratories dairy farm.

The first line of the Abstract of the 1930 Cow Milking Apparatus (Rotolactor) patent starts:

> "The object of this invention is to provide an apparatus whereby an indefinitely large number of cows may be milked successively and largely automatically."

The Rotolactor (*roto* + *lactor* ium) was a large rotating "merry-go-round" style platform for holding 50 cows. The machine brought the cows into position for milking with automatic milking machines. The rotating platform machine was sixty feet in diameter and made one complete revolution about every twelve and a half minutes, which was the time required to prepare and milk each cow.

The first step for each new cow was receiving a bath. They were bathed with warm water and automatic showers, supplemented by two men using pressure hoses, who "devote their attention to the cleansing of udders and flanks."

The next operator prepared the udder for the milking. Then the teat-cups of the automatic milking machine were attached to the cow's udder. The cow was then milked for the twelve and a half minutes during the Rotolactor's one-time complete rotation. The teat-cups would then be detached at the end of the twelve and a half minute rotation. The cow would then step off the platform and return to the barn to her stall.

The milk was drawn by a vacuum to sealed glass containers above the cow's head. It was then transported in pipes to weighing and recording apparatuses. Then it was piped to another room where the milk was cooled and bottled. This was faster and more efficient than the methods previously used. Human hands never touched the milk, and the milk never came into contact with air, which was important to prevent premature spoiling. The Rotolactor could milk the Walker-Gordon dairy's 1,680 cows three times daily. This produced 26,000 quarts of milk.

Bender Machine Works

The Bender Machine Works, is a dairy equipment manufacturer that played a major role in the history of the dairy farming business in the United States from the 1950s to the 1980s, producing milk pipeline and milk transfer cart components, and washing/vacuum-releasing equipment.

The company continues to be in business but now focuses its manufacturing on private label electromechanical milk pipeline washing systems.

Washer/Releaser and Step-saver

The Bender Washer/Releaser was a vacuum-operated device first invented and patented by Lloyd Bender of Hayward, WI, USA in the 1950s, used to both wash dairy milking equipment and to transfer milk from piping containing a vacuum, into a storage tank at normal atmospheric pressure.

As a non-electric pneumatically operated device, the washer/releaser could be used on small dairy farms without electricity that used an engine to supply milking vacuum, and used well water to cool milk.

The Washer/Releaser could be used in association with both dairy pipelines and the *Step-Saver*, a milk transfer cart used with bucket milkers. The Step-Saver and the Washer/Releaser were once extremely common in the 1960s to 1980s on small family dairy farms in the United States, to reduce the labor of bucket milking before the milk pipeline became more popular.

Purpose and Usage

As barns began to increase in size from perhaps 6 to 12 cows to 30 or 40 cows, the bucket milker became a very laborious milking system. As the barn length increased, the farmer had to walk an increasing distance from the cow to the milk bulk tank to dump the collected milk. An early vacuum milk-transport system known as the Step-Saver was developed to save the farmer the trouble of carrying the heavy steel buckets of milk all the way back to the storage tank in the milkhouse. The system used a very long vacuum hose coiled around a receiver cart, and connected to a vacuum-breaker device in the milkhouse.

Following milking each cow, the bucket milker would be dumped into the receiver cart. A foot pedal on the base of the cart lifted the cover, which kept contaminating dust and debris out of the cart, and allowed the farmer to hold the heavy bucket milker with both hands while pouring. A diffuser plate in the top of the cart prevented milk from splashing out while rapidly pouring the milk, and a large filter disk under the diffuser removed any debris from the milk.

Milk collected in a chamber below the filter, and was slowly sucked through the long hose to the milkhouse. When empty, a large float ball in the bottom of the cart would settle down over the drain hole to seal the line and retain system vacuum. When milk was poured into the cart, the ball would float up, unsealing the drain.

An automatic vacuum breaker in the milkhouse cyclically pulled milk from the cart into a glass jar using system vacuum, followed by a release of vacuum to atmospheric pressure, allowing the milk to flow into the bulk tank by gravity flow. When the float level in the jar dropped to setpoint, system vacuum was reapplied to restart the process. Check valves on the vacuum breaker milk hose prevented milk from flowing backwards to the cart when the jar vacuum was released.

As the farmer milked the cows in series, the cart would be rolled further down the center aisle, the long milk hose unwrapped from the cart, and hung on open hooks along the ceiling of the aisle.

Wisconsin Legislative Code References

Use of the step saver and similar non-pipeline milk transfer systems is described in the legislative code of the Wisconsin Department of Agriculture, Trade, and Consumer Protection (DATCP).

Milking and Milk Handling Systems

Non–pipeline Systems: If milk from milking animals is initially collected in a portable transfer receptacle and pumped to the milkhouse through a flexible tube, rather than being pumped directly to the milkhouse through a permanently mounted pipeline, the transfer receptacle and tube system shall comply with the following requirements:

- The portable transfer receptacle shall be constructed of stainless steel or an equally corrosion resistant metal, and shall have an overlapping self–closing cover. The receptacle shall be supported off the floor on a cart or mobile structure which can be easily cleaned.

• The tube used to transfer milk from the portable transfer receptacle to the milkhouse shall consist of a single length of transparent tubing material. The milk transfer tube shall be supported off the floor at all times. The interior milk contact surface of the transfer tube shall be mechanically cleaned, sanitized and dried after each use. The opening through which the milk transfer tube enters the milkhouse shall be kept closed when the tube is not in use. A milk transfer tube shall not be left suspended in a milking barn or parlor between uses, but shall be stored in the milkhouse.

Electromechanical Pipeline Washing

The Bender Machine Works also developed some of the first automated "wash-in-place" milk pipeline systems, which is generally constructed similar to an automatic dishwasher with automated fill, drain, soap dispensing, and various wash cycles. They continue to manufacture these washing systems today.

Bulk Milk Cooling Tank

In dairy farming a bulk milk cooling tank is a large storage tank for cooling and holding milk at a cold temperature until it can be picked up by a milk hauler. The bulk milk cooling tank is an important piece of dairy farm equipment. It is usually made of stainless steel and used every day to store the raw milk on the farm in good condition. It must be cleaned after each milk collection. The milk cooling tank can be the property of the farmer or be rented from a dairy plant.

Bulk Tank Types

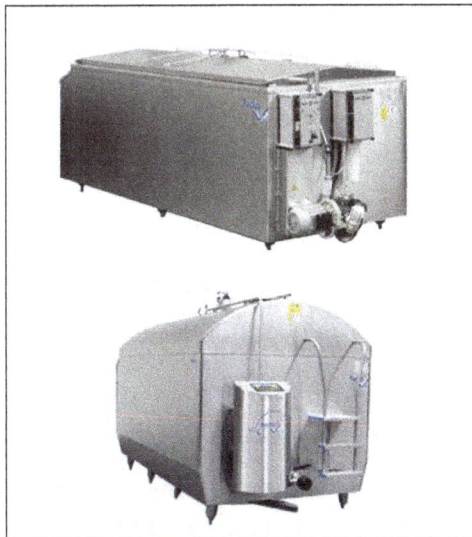

Different types of milk cooling tanks.

Raw milk producers have a choice of either open (from 150 to 3000 litres) or closed (from 1000 to 10000 litres) tanks. The cost can vary considerably, depending on manufacturing norms and whether a new or second hand tank is purchased.

Milk silos (10,000 litres and plus) are suitable for the very large producer. These are designed to be installed outside and adjacent to the dairy, all controls and the milk outlet pipe being situated in the dairy.

Tank Construction

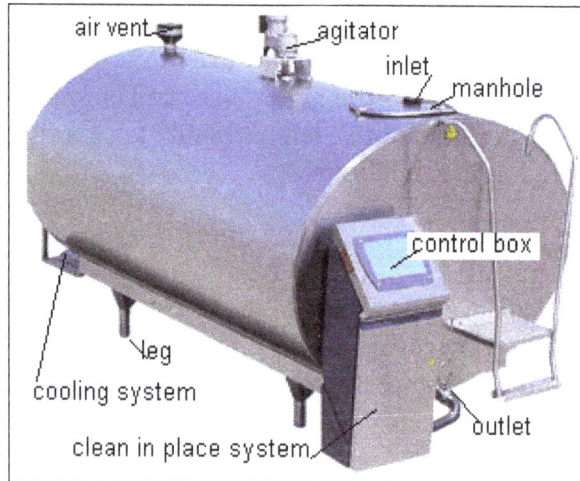

Bulk milk cooling tank.

A milk cooling tank, also known as a bulk tank or milk cooler, consists of an inner and an outer tank, both made of high quality stainless steel.

The space between the outer tank and the inner tank is isolated with polyurethane foam. In case of a power failure with an outside temperature of 30 °C, the content of the tank will warm up only 1 °C in 24 hours.

To facilitate an adequate and rapid cooling of the entire content of a tank, every tank is equipped with at least one agitator. Stirring the milk ensures that all milk inside the tank is of the same temperature and that the milk stays homogeneous.

On top of every closed milk cooling tank is a manhole of about 40 centimetres diameter. This enables thorough cleaning and inspection of the inner tank if necessary. The manhole is covered by a lid and sealed watertight with a rubber ring. Also on top are 2 or 3 small inlets. One is covered with an air-vent, the other(s) can be used to pump milk into the tank.

A milk cooling tank usually stands on 4, 6, or 8 adjustable legs. The built-in tilt of the inner tank ensures that even the last drop of milk will eventually flow to the outlet.

At the bottom, every milk cooling tank has a threaded outlet, usually including a valve. All tanks have a thermometer, allowing for immediate inspection of the inner temperature.

Most tanks include an automatic cleaning system. Using hot and cold water, an acid and alkaline cleaning fluid, a pump and a spray lance will clean the inner tank, ensuring an hygienic inner environment each time the tank is emptied.

Almost every tank has a control box. It manages the cooling process by use of a thermostat. The user can turn the system on and off, allow for extra and immediate stirring, start the cleaning routine, and reset the entire system in case of a failure.

New and bigger milk cooling tanks are now being equipped with monitoring and alarm systems. These systems guard temperature of the milk inside the tank, check the functioning of the agitator, the cooling unit and temperature of the cleaning water. In case of malfunctioning of any of these functions, the alarm will activate. The monitoring system will also keep a record of the temperature and of all malfunctions for a given period.

Bulk Tank Manufacturing Norms

Norms define among other criteria: insulation, milk agitation, cooling power required, variations in milk quantity measurement, calibration, etc. Some are more demanding than others.

- ISO standard 5708 (Refrigerated bulk milk tanks).

- European standard EN 13732 (Food processing machinery – Bulk milk coolers on farms – Requirements for construction, performance, suitability for use, safety and hygiene).

- Northern American sanitary standard 3A 13-11.

Bulk Tank Outlet Standards

Swedish outlet (SMS 1145), German outlet (DIN 11851), English RJT (BS 4825), IDF (ISO 2853), tri-clamp (ISO 2852), Danish outlet (DS 722), can be found, not to mention different diameters. They vary from country to country. Non standard outlets make the milk collection process difficult, as the operator needs to adapt to each different standard/diameter.

Cooling Systems

There are two primary methods of cooling milk entering the bulk tank, each with its own advantages and disadvantages. The tank capacity and type will depend on herd size, calving pattern, frequency of milk collection, required milk quality, energy and water availability and future plans for development.

Direct Expansion

A bulk tank with direct expansion cooling has pipes or pillow plates carrying refrigerant which are welded directly to the exterior of the milk chamber. A layer of insulation covers the exterior of the milk tank and the cooling lines, with an exterior metal shell over the insulation.

Direct expansion cooling cannot run when the tank is empty or the inside walls of the tank would freeze. Instead, the tank is rapidly cooled as warm milk first enters the tank, and then the tank is cooled slowly just to maintain a low storage temperature. The rapid cooling during milking requires very large refrigeration compressors and condenser radiators to quickly expel heat from the milk, and is better suited for very large farming operations where three-phase electric power is available to operate the high-power cooling system.

Ice Bank

A bulk tank using an Ice Builder or Ice Bank immerses the bottom of the inner milk chamber in an open pool of water with copper tubes containing refrigerant suspended in the water. Between milkings, a small low-power cooling system slowly builds up a coating of ice around the copper tubes, and prevents icing of the pool over by continuously circulating the water in the pool. After the ice has achieved a thickness of 2-3 inches, the cooling system stops running.

During milking, the milk entering the tank is primarily cooled by circulating the water in the pool around the walls of the inner milk chamber, and the melting of the ice. After the ice has melted sufficiently the cooling system restarts to assist the ice bank and restart the ice building.

Ice bank bulk tanks are better suited for small family farm operations where only single-phase electric power is available, and high-power cooling systems would be either too expensive or difficult to install.

Milk Pre-cooling

For energy savings and quality reasons it is advisable to pre-cool the milk before it enters the tank using a plate or a tube cooler (shell and tube heat exchanger) supplied with chilled water from the well water, the ice builder or the condensing unit. The quicker milk is cooled after leaving the cow the better. This system achieves most of the cooling before the milk enters the tank, so that chilled milk, rather than warm milk, is being added to the already cooled milk in the tank.

Cooling Temperature

Generic temperature for milk storage is 3 to 4 °C. For raw milk cheese manufacturing, it would be advisable to keep the milk at 12 °C, as milk characteristics will be kept in a better state.

The milk cooling tank is usually not completely filled at once. A 2 milking tank is designed to cool 50% of its capacity at once. A 4 milking tank is designed to cool 25% of its capacity at once, and a 6 milking tank is designed to cool 16.7% of its capacity at once.

The cooling performance depends on the number of milking it takes to completely fill the tank, the ambient temperature and the cooling time.

Bulk Tank Cleaning Systems

There are two primary methods of cleaning bulk tanks, via manual scrubbing or automatic washing. Both methods generally use four steps to clean the tank:

- Prerinsing with water to wet the surface and rinse off remaining milk residue.
- Washing with hot soapy water.
- Rinsing with water to remove the soap.
- Final sanitizing rinse with an approved bulk tank sanitizer solution.

Manual Scrubbing

Manual scrubbing requires the bulk tank to have large hinged covers that can be lifted open to permit easy access to the interior surfaces of the tank. It tends to be much more thorough than automatic methods since it permits the tank to be carefully inspected during the washing process.

If the tank is not found to be cleaned well enough, a troublesome area can be given additional cleansing attention.

Manual Scrubbing Limitations

This job is difficult to perform for very large tanks, and becomes more difficult as the overall cross-section or diameter of the tank increases, requiring either a longer brush or a raised work platform around the tank to lift the cleaning worker to reach over the side of a tall tank.

Automatic Washing

Automatic bulk tank washing and are normally activated by the milk collection truck driver after each milk collection. The cleaning system operates similar to a consumer dishwasher and consists of one or more free-spinning high-pressure spray nozzles with tangential jets, with the spray nozzle mounted on the end of a flexible whip suspended down into the center of the interior. As the cleaning solution sprays out of the jet, the force of the expelled water causes the jet to spin around and the whip to wildly swing back and forth, spraying the cleaning solution randomly all over the interior of the tank.

Automatic Washing Limitations

Because no physical scrubbing occurs with automatic wash systems, the cleanser relies on surfactants and detergents to dissolve the fats left on the interior of the tank by the cream in the milk. However, this is not sufficient to remove milkstone buildup, and the tank may need to be washed occasionally with milkstone remover to remove this scale buildup that can harbor bacteria and contaminants.

Automatic scrubbing only cleans the interior of the tank. It is not capable of cleaning the exterior of the tank, and it does not do a good job of washing around the cover seals. While it is possible to just clean the interior and call it good enough, it does not provide the maximum sanitation of manually washing down the exterior of the tank following or during the automatic wash process. Also, some components that contact the milk such as the drain valve cannot be properly cleaned automatically without disassembling the valve and retaining washer and directly scrubbing in soapy water.

Operating Costs

Substantial reductions in running costs can be made when an ice builder is used in conjunction with off-peak electricity. Pre-cooling milk using a plate or a tube cooler supplied with mains or well water can also reduce costs and add to the cooling capacity of the tank.

Bulk tank condenser units, which are not an integral part of the tank, should be fitted in an adjacent, suitable and well ventilated place. If at all possible, condenser units should not be fitted on a wall facing the sun. They should be installed in a way which allows them to draw in and discharge adequate quantities of air for efficient operation.

Bulk tank should be easily accessible by large bulk collection tankers and positioned so that the tanker approaches can be kept clean and free from cow traffic at all times.

Although tanks have been calibrated when first installed, bulk tank miscalibration is not uncommon and in some cases it can result in significant loss of income. Milk tanks calibrated on the low side, can cheat raw milk producers by up to 22 litres on each shipment. It is therefore advisable to re-calibrate a bulk tank.

Other Usage of Bulk Tanks

Stainless steel bulk tanks are also used to heat or cool a fluid or simply to keep it isolated and warm/cold. Because of the hygienical finishing of the inner and outer side of the tanks, almost any fluid can be stored: water, fruit juices, honey, wine, beer, ink, paint, cosmetics, aromatic food-additives, bacterial cultures, cleansers, oil, or blood.

Mixer-wagon

A paddle-type mixer-wagon
coupled to a tractor.

A weighing computer showing an
empty mixer-wagon, 0 kg tare.

A mixer-wagon, or diet feeder, is a specialist agricultural machine used for accurately weighing, mixing and distributing total mixed ration (TMR) for ruminant farm animals, in particular cattle and most commonly, dairy cattle.

Trailed mixer-wagons vary in size from 5 m³ to more than 45 m³. Some self-propelled mixer-wagons may be bigger than this. Displacement varies according to ration dry matter. More water means more weight. With dry (45% DM; dry matter) rations, a 14 m³ mixer-wagon such as the one pictured, may contain 3 tonnes fully loaded, or enough for about 60 Holstein cows.

A mixer-wagon commonly consists of:

- A trailed chassis, for coupling to a tractor power unit, and fitted with one or more usually braked axles, and fitted with a road-legal lighting system.

- A mixing body, attached to the chassis by four weighing sensors, one at each corner. There are three main types of mixing body:

  ◦ Paddle, whereby a central axial shaft turns a series of paddles which rotate the contents and provide front to back mixing in the mixer-wagon.

  ◦ Vertical auger, with one, two or three augers used to move the contents from top to bottom.

- ◦ Horizontal auger, containing between one and five augers, used to circulate the contents from front to back and bottom to top.

- A digital weighing computer working through the above-mentioned weighing sensors. Such a computer can typically memorise 9 or more different rations and 99 or more ingredients.

The paddle turning inside a mixer-wagon. Stationery knives are fitted on the side of the tub.

The system can weigh in all the ingredients of a ration in any chosen order and weigh out according to chosen quantities corresponding to the needs of different groups of animals. As each ingredient is loaded, a visual and audible signal alerts the operator when the required amount is reached. The flashing light and "beep" system repeats faster and faster until the ingredient is complete, when the signal becomes continuous for 2 seconds. The computer then shifts to the next ingredient and displays the product name and weight to be loaded. Loading is generally done using the mixer-wagon cutting and loading device, a high capacity tractor loader, or a telescopic handler.

- The mixing paddle, rotors or augers are connected to the tractor pto through a reducer system, provided by either a planetary gearbox, and a step-down pulley and chain system.

- A set of stationary knives, against which long fibre may be chopped by the forcing action of the mixing rotor.

- A hydraulic door to seal the ration in during mixing, thus permitting the use of liquid feeds such as molasses.

- An unloading system, consisting of a simple hydraulically adjustable chute, up to a hydraulic powered conveyor belt.

- Self-propelled mixer-wagons are mounted on a lorry chassis or may be specialist self-loading machines.

Automatic Milking

Automatic milking is the milking of dairy animals, especially of dairy cattle, without human labour. Automatic milking systems (AMS), also called voluntary milking systems (VMS), were developed in the late 20th century. They have been commercially available since the early 1990s. The core of such systems that allows complete automation of the milking process is a type of agricultural robot.

Automated milking is therefore also called robotic milking. Common systems rely on the use of computers and special herd management software. Also it used to monitor the health status of cows.

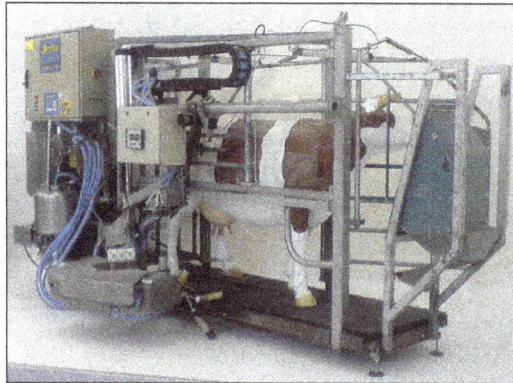

A Fullwood Merlin AMS unit from the 1990s, exhibit at the Deutsches Museum in Germany.

Automated Milking

Milking Process and Milking Schedules

The milking process is the collection of tasks specifically devoted to extracting milk from an animal (rather than the broader field of dairy animal husbandry). This process may be broken down into several sub-tasks: collecting animals before milking, routing animals into the parlour, inspection and cleaning of teats, attachment of milking equipment to teats, and often massaging the back of the udder to relieve any held back milk, extraction of milk, removal of milking equipment, routing of animals out of the parlour.

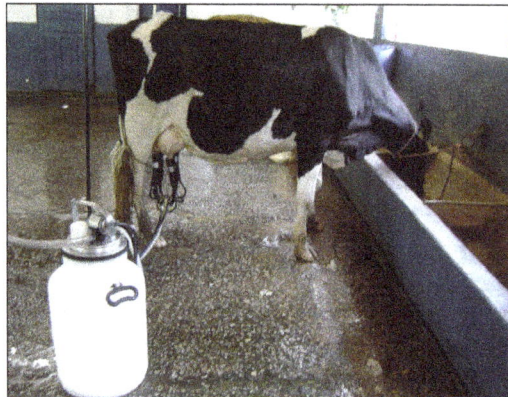

A cow and a milking machine – partial automation compared to hand milking.

Maintaining milk yield during the lactation period (approximately 300 days) requires consistent milking intervals, usually twice daily and with maximum time spacing between milkings. In fact all activities must be scheduled around the milking process on the dairy farm. Such a milking routine imposes restrictions on time management and personal life of an individual farmer, as the farmer is committed to milking in the early morning and in the evening for seven days a week regardless of personal health, family responsibilities or social schedule. This time restriction is exacerbated for lone farmers and farm families if extra labour cannot easily or economically be obtained, and is a factor in the decline in small-scale dairy farming. Techniques such as once-a-day milking and voluntary milking have been investigated to reduce these time constraints.

A rotary milking parlor – higher efficiency compared to stationary milking parlors, but still requiring manual labour with milking machines etc.

Automation Progress in the 20th Century

To alleviate the labour involved in milking, much of the milking process has been automated during the 20th century: many farmers use semi-automatic or automatic cow traffic control (powered gates, etc.), the milking machine (a basic form was developed in the late 19th century) has entirely automated milk extraction, and automatic cluster removal is available to remove milking equipment after milking. Automatic teat spraying systems are available, however, there is some debate over the cleaning effectiveness of these.

The final manual labour tasks remaining in the milking process were cleaning and inspection of teats and attachment of milking equipment (milking cups) to teats. Automatic cleaning and attachment of milking cups is a complex task, requiring accurate detection of teat position and a dextrous mechanical manipulator. These tasks have been automated successfully in the voluntary milking system (VMS), or automatic milking system (AMS).

Automatic Milking Systems

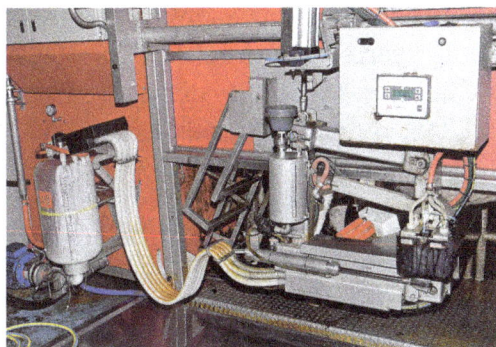

An older Lely Astronaut AMS unit at work (milking).

Since the 1970s, much research effort has been expended in investigating methods to alleviate time management constraints in conventional dairy farming, culminating in the development of the automated voluntary milking system.

Voluntary milking allows the cow to decide her own milking time and interval, rather than being milked as part of a group at set milking times. AMS requires complete automation of the milking process as the cow may elect to be milked at any time during a 24-hour period.

The milking unit comprises a milking machine, a teat position sensor (usually a laser), a robotic arm for automatic teat-cup application and removal, and a gate system for controlling cow traffic. The cows may be permanently housed in a barn, and spend most of their time resting or feeding in the free-stall area. If cows are to be grazed as well, using a selection gate to allow only those cows that have been milked to the outside pastures has been advised by some AMS manufacturers.

When the cow elects to enter the milking unit (due to highly palatable feed that she finds in the milking box), a cow ID sensor reads an identification tag (transponder) on the cow and passes the cow ID to the control system. If the cow has been milked too recently, the automatic gate system sends the cow out of the unit. If the cow may be milked, automatic teat cleaning, milking cup application, milking, and teatspraying takes place. As an incentive to attend the milking unit, concentrated feedstuffs needs to be fed to the cow in the milking unit.

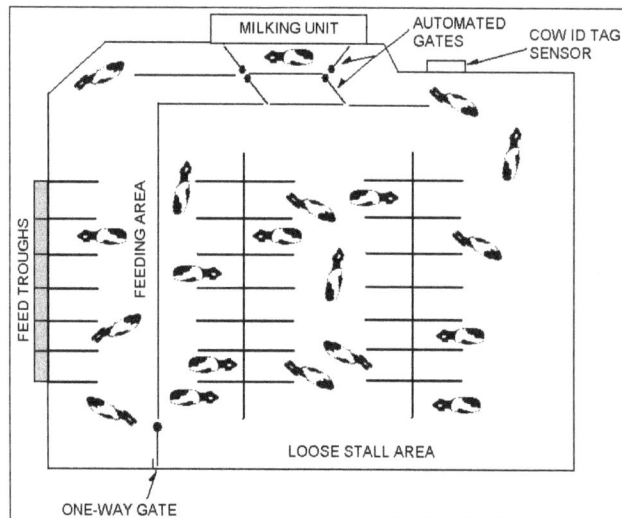

Typical VMS stall layout (forced cow traffic layout).

The barn may be arranged such that access to the main feeding area can only be obtained by passing the milking unit. This layout is referred to as forced cow traffic. Alternatively, the barn may be set up such that the cow always has access to feed, water, and a comfortable place to lie down, and is only motivated to visit the milking system by the palatable feed available there. This is referred to as free cow traffic.

The innovative core of the AMS system is the robotic manipulator in the milking unit. This robotic arm automates the tasks of teat cleaning and milking attachment and removes the final elements of manual labour from the milking process. Careful design of the robot arm and associated sensors and controls allows robust unsupervised performance, such that the farmer is only required to attend the cows for condition inspection and when a cow has not attended for milking.

Typical capacity for an AMS is 50–70 cows per milking unit. AMS usually achieve milking frequencies between 2 and 3 times per day, so a single milking unit handling 60 cows and milking each cow 3 times per day has a capacity of 7.5 cows per hour. This low capacity is convenient for lower-cost design of the robot arm and associated control system, as a window of several minutes is available for each cow and high-speed operation is not required.

AMS units have been available commercially since the early 1990s, and have proved relatively successful in implementing the voluntary milking method. Many of the research and developments have taken place in the Netherlands. The most farms with AMS are located in the Netherlands, and Denmark.

A new variation on the theme of robotic milking includes a similar robotic arm system, but coupled with a rotary platform, improving the number of cows that can be handled per robot arm. A mobile variation of robotic milking, adapted to tie-stall configuration (stanchion barns), is used in Canada. In this configuration, the AMS travels in the centre isle of the barn approaching cows from behind to milk them in their stalls.

Advantages

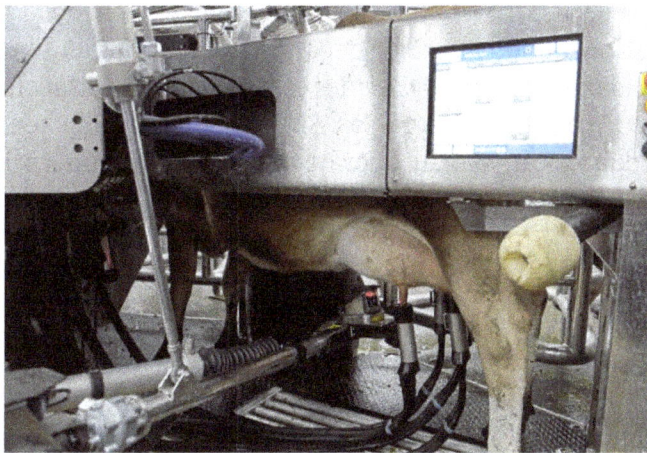

An AMS unit at work (teat cleaning).

- Elimination of labour: The farmer is freed from the milking process and associated rigid schedule, and labour is devoted to supervision of animals, feeding, etc.

- Milking consistency: The milking process is consistent for every cow and every visit, and is not influenced by different persons milking the cows. The four separate milking cups are removed individually, meaning that an empty quarter does not stay attached while the other three are finishing, resulting in less threat of injury. The newest models of automatic milkers can vary the pulsation rate and vacuum level based on milk flow from each quarter.

- Increased milking frequency: Milking frequency may increase to three times per day, however typically 2.5 times per day is achieved. This may result in less stress on the udder and increased comfort for the cow, as on average less milk is stored. Higher frequency milking increases milk yield per cow, however much of this increase is water rather than solids.

- Perceived lower stress environment: There is a perception that elective milking schedules reduce cow stress.

- Herd management: The use of computer control allows greater scope for data collection. Such data allows the farmer to improve management through analysis of trends in the herd, for example response of milk production to changes in feedstuffs. Individual cow histories may also be examined, and alerts set to warn the farmer of unusual changes indicating illness or injury. Information gathering provides added value for AMS, however correct interpretation and use of such information is highly dependent on the skills of the user or the accuracy of computer algorithms to create attention reports.

Considerations and Disadvantages

- Higher initial cost: AMS systems cost approximately €120,000 ($190,524) per milking unit as of 2003 (presuming barn space is already available for loose-stall housing). Equipment costs decreased from $175,000 for the first stall to $158,000. Equipment costs decreased from $10,000/stall for a double-six parlor to $9000/stall for a double-ten parlor with a cost of $1200/stall for pipeline milking. Initial parlor cost was increased $5000/stall to represent a high cost parlor. Whether it is economically beneficial to invest in an AMS instead of a conventional milking parlor depends on constructions costs, investments in the milking system and costs of labour. Besides costs of labour, the availability of labour should also be taken into account. In general, an AMS is economically beneficial for smaller scale farms, and large dairies can usually operate more cheaply with a milking parlor.

- Increased electricity costs: To operate the robots, but this can be more than outweighed by reduced labour input.

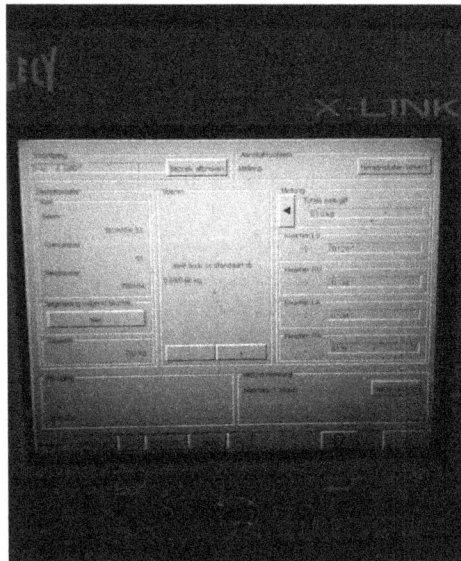

Touchscreen display of a milking robot.

- Increased complexity: While complexity of equipment is a necessary part of technological advancement, the increased complexity of the AMS milking unit over conventional systems, increases the reliance on manufacturer maintenance services and possibly increasing operating costs. The farmer is exposed in the event of total system failure, relying on prompt response from the service provider. In practice AMS systems have proved robust and manufacturers provide good service networks. Because all milking cows have to visit the AMS voluntarily, the system requires a high quality of management. The system also involves a central place for the computer in the daily working routines.

- Difficult to apply in pasture systems: As a continuous animal is pediment for an optimal utilization of the AMS unit, AMS works at their best in zero-grazing systems, in which the cow is housed indoors for most of the lactation period. Zero-grazing suits areas (e.g. the Netherlands) where land is at a premium, as maximum land can be devoted to feed production which is then collected by the farmer and brought to the animals in the barn. In pasture systems, cows graze in fields and are required to walk to the milking parlour. It has been found that it can be challenging to make cows maintain a high milking frequency if the distance to walk between pasture and milking unit is too great. Maintaining production on pasture has, however, been shown to be possible in amongst others the autograssmilk project. where cattle are on pasture and milked by AMS.

- Lower milk quality: With automatic milking, the number of anaerobic spores, the freezing point increases, the frequency of milk quality failure almost doubles, which fully reflects the quality of milk caused by automatic milking. Although the automatic milking machine cleans the cow's teat and tests the pre-squeezed milk, there is still a phenomenon that the infected milk is not transferred, and the device itself is also lack of cleaning, and the milk is not handled properly. This situation was also confirmed in 2002 when investigating nearly 98 farms in Denmark with automatic milking systems. Bulk milk total bacteria count (BMTBC) and somatic cell count (BMSCC) are also affected by automatic milking. These two counts were studied when introducing an automatic milking system to cows that were previously milked conventionally. BMSCC was found to not significantly increase between pre and post-AMS installation but BMTBC was found to significantly increase in the first three months but then return to normal levels. BMSCC was found to significantly improve in the third year with respect to the pre-introduction level.

- Possible increase in stress for some cows: Cows are social animals, and it has been found that due to dominance of some cows, others will be forced to milk only at night. Such behavior is inconsistent with the perception that AM reduces stress by allowing "free choice" of milking time.

- Decreased contact between farmer and herd: Effective animal husbandry requires that the farmer be fully aware of herd condition. In conventional milking, the cows are observed before milking equipment is attached, and ill or injured cows can be earmarked for attention. Automatic milking decreases the time the farmer has such close contact with the animal, with the possibility that illness may go unnoticed for longer periods and both milk quality and cow welfare suffer. In practice, milk quality sensors at the milking unit attempt to detect changes in milk due to infection, and farmers inspect the herd frequently (Farmers still need to provide bedding for the cows, provide reproductive health services, feed them, and occasionally repair parts of the barn). However this concern has meant that farmers are still tied to a seven-day schedule. Modern automatic milking systems attempt to rectify this problem by gathering data that would not be available in many conventional systems including milk temperature, milk conductivity, milk color including infrared scan, change in milking speed, change in milking time or milk letdown by quarter, cow's weight, cow's activity (movements), time spent ruminating, etc.

- Dependence on the robotics company: The maintenance becomes significantly more time critical and may put the farmer at greater risk. For example, one farm in Estonia reported losses of over 1 million euros, when the robots from BouMatic Robotics performed promised standards and the company failed to provide maintenance.

Milk Pipeline

Milk pipeline uses a permanent milk-return pipe and a second vacuum pipe that encircles the barn or milking parlor above the rows of cows, with quick-seal entry ports above each cow. By eliminating the need for the milk container, the milking device shrank in size and weight to the point where it could hang under the cow, held up only by the sucking force of the milker nipples on the cow's udder. The milk is pulled up into the milk-return pipe by the vacuum system, and then flows by gravity to the milkhouse vacuum-breaker that puts the milk in the storage tank. The pipeline system greatly reduced the physical labor of milking since the farmer no longer needed to carry around huge heavy buckets of milk from each cow.

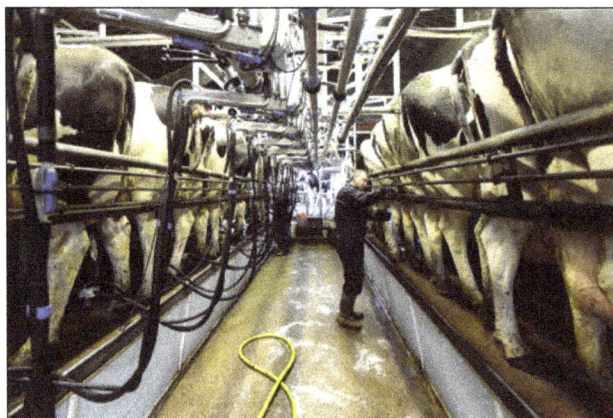

The pipeline allowed barn length to keep increasing and expanding, but after a point farmers started to milk the cows in large groups, filling the barn with one-half to one-third of the herd, milking the animals, and then emptying and refilling the barn. As herd sizes continued to increase, this evolved into the more efficient milking parlor.

Milking Parlors

Innovation in milking focused on mechanising the milking parlour to maximise throughput of cows per operator which streamlined the milking process to permit cows to be milked as if on an assembly line, and to reduce physical stresses on the farmer by putting the cows on a platform slightly above the person milking the cows to eliminate having to constantly bend over. Many older and smaller farms still have tie-stall or stanchion barns, but worldwide a majority of commercial farms have parlours.

The milking parlor allowed a concentration of money into a small area, so that more technical monitoring and measuring equipment could be devoted to each milking station in the parlor. Rather than simply milking into a common pipeline, for example, the parlor can be equipped with fixed measurement systems that monitor milk volume and record milking statistics for each animal. Tags on the animals allow the parlor system to automatically identify each animal as it enters the parlor.

Recessed Parlors

More modern farms use recessed parlors, where the milker stands in a recess such that his arms are at the level of the cow's udder. Recessed parlors can be herringbone, where the cows stand in two angled rows either side of the recess and the milker accesses the udder from the side, parallel, where the cows stand side-by-side and the milker accesses the udder from the rear or, more recently, rotary (or carousel), where the cows are on a raised circular platform, facing the center of the circle, and the platform rotates while the milker stands in one place and accesses the udder from the rear. There are many other styles of milking parlors which are less common.

Herringbone and Parallel Parlors

In herringbone and parallel parlors, the milker generally milks one row at a time. The milker will move a row of cows from the holding yard into the milking parlor, and milk each cow in that row. Once all or most of the milking machines have been removed from the milked row, the milker releases the cows to their feed. A new group of cows is then loaded into the now vacant side and the process repeats until all cows are milked. Depending on the size of the milking parlor, which normally is the bottleneck, these rows of cows can range from four to sixty at a time.

Rotary Parlors

In rotary parlors, the cows are loaded one at a time onto the platform as it slowly rotates. The milker stands near the entry to the parlor and puts the cups on the cows as they move past. By the time the platform has completed almost a full rotation, another milker or a machine removes the cups and the cow steps backwards off the platform and then walks to its feed.

Automatic Milker Take-off

It can be harmful to an animal for it to be over-milked past the point where the udder has stopped releasing milk. Consequently the milking process involves not just applying the milker, but also monitoring the process to determine when the animal has been milked out and the milker should be removed. While parlor operations allowed a farmer to milk many more animals much more quickly, it also increased the number of animals to be monitored simultaneously by the farmer. The automatic take-off system was developed to remove the milker from the cow when the milk flow reaches a preset level, relieving the farmer of the duties of carefully watching over 20 or more animals being milked at the same time.

Fully Automated Robotic Milking

In the 1980s and 1990s, robotic milking systems were developed and introduced (principally in the EU). Thousands of these systems are now in routine operation. In these systems the cow has a high degree of autonomy to choose her time of milking within pre-defined windows. These systems are generally limited to intensively managed systems although research continues to match them to the requirements of grazing cattle and to develop sensors to detect animal health and fertility automatically.

References

- Centre, Government of Canada;Canadian Dairy Information. "Dairy Animal Registrations – Canadian Dairy Information Centre (CDIC)". Www.dairyinfo.gc.ca. Retrieved 29 January 2018

- What-is-a-dairy-farm, pages: extension.org, Retrieved 22 June, 2019

- Knaus (2009). "Dairy cows trapped between performance demands and adaptability". Journal of the Science of Food and Agriculture. 89 (7): 1107–1114. Doi:10.1002/jsfa.3575

- Dairy-farm-management-scope-and-importance, essay: biologydiscussion.com, Retrieved 23 July, 2019

- Devries, & Von Keyserlingk. (2005). Time of Feed Delivery Affects the Feeding and Lying Patterns of Dairy Cows. Journal of Dairy Science, 88(2), 625–631

- Managing-cow-lactation-cycles: thecattlesite.com, Retrieved 24 August, 2019

- Castro, Angel (2018). "Long-term variability of bulk milk somatic cell and bacterial counts associated with dairy farms moving from conventional to automatic milking systems". Italian Journal of Animal Science. Journal of Animal Science. 17: 218–225. Doi:10.1080/1828051X.2017.1332498

- Currently-available-precision-dairy-farming-technologies, dairy: afs.ca.uky.edu, Retrieved 25 January, 2019

- Hopster, H., et al., (2002), "Stress Responses during Milking; Comparing Conventional and Automatic Milking in Primiparous Dairy Cows", Journal of Dairy Science Vol. 85, pp. 3206–3216

- Milking-pipeline: farmdairy.blogspot.com

Farm Equipments and Tools

Some of the tools and equipments which are used in farming are tractors, seed drills, harrows, rollers, cultivators, mowers, balers, hay rakes, tedders and combine harvesters. This chapter has been carefully written to provide an easy understanding of the varied applications of these equipments and tools in farming.

FARM MACHINERY

Farm equipment is any kind of machinery used on a farm to help with farming. The best-known example is a tractor. There are also many other farm implement used since prehistoric times to the complex harvesters of modern mechanized agriculture.

The operations of farming for which machines are used are diverse. For crop production they include handling of residues from previous crops; primary and secondary tillage of the soil; fertilizer distribution and application; seeding, planting, and transplanting; cultivation; pest control; harvesting; transportation; storage; premarketing processing; drainage; irrigation and erosion control; and water conservation. Livestock production, which not so long ago depended primarily on the pitchfork and scoop shovel, now uses many complicated and highly sophisticated machines for handling water, feed, bedding, and manure, as well as for the many special operations involved in producing milk and eggs.

In the early 19th century, animals were the chief source of power in farming. Later in the century, steam power gained in importance. During World War I gasoline- (petrol-) powered tractors became common, and diesel engines later became prevalent. In the developed countries, the number of farm workers has steadily declined in the 20th century, while farm production has increased because of the use of machinery.

TRACTOR

Tractors have traditionally been used on farms to mechanise several agricultural tasks. Modern tractors are used for ploughing, tilling and planting fields in addition to routine lawn care, landscape maintenance, moving or spreading fertiliser and clearing bushes.

Tractors offer advantages on small farms as well as in regular lawn and garden work.

Wide Range

Tractors are available in a wide range of options to suit specific tasks and requirements. Subcompact or compact tractors available in a horsepower range of 15 hp to 40 hp are ideal for heavy duty landscaping jobs and tasks such as digging, hauling or ploughing on large gardens, fields and pastures.

A smaller version of compact tractors, subcompact tractors have the power and versatility to perform a large range of gardening tasks including mowing, moving mulch and tilling gardens. Compact tractors are a smaller version of utility tractors and are ideal for landscaping tasks.

Also known as diesel tractors, utility tractors are recommended for mechanising complex farming tasks and come in different models with a horsepower range of 45 hp to 110 hp. A wide range of farming implements can be attached to utility tractors to help accomplish various jobs.

Versatility

Though available in a wide range of models, modern tractors are designed and manufactured to offer versatility in performing a wide range of tasks. Compact tractors can accomplish tasks ranging from gardening to simple farming jobs, with the ability to do more by attaching various implements such as front loaders or back hoes.

Power and Durability

Tractors are typically designed with powerful engines to run over rough terrain and pull extremely heavy loads, making them effective in tough farming or landscape tasks. Modern tractors also come with cast iron front axles for extra strength and durability.

Ease of Transmission and Operation

Modern tractors feature powershift transmission and hydrostatic transmission to simplify operation. While these tractors are also provided with power steering to make turning much easier, advanced models help reduce operator fatigue with exclusive shift controls and an automatically responsive transmission.

Planter

A two row planter featuring John Deere "71 Flexi" row units.

A planter is a farm implement, usually towed behind a tractor, that sows (plants) seeds in rows throughout a field. It is connected to the tractor with a drawbar or a three-point hitch. Planters lay the seeds down in precise manner along rows. Planters vary greatly in size, from 1 row to 54, with the biggest in the world being the 48-row John Deere DB120. Such larger and newer planters comprise multiple modules called row units. The row units are spaced evenly along the planter at intervals that vary widely by crop and locale. The most common row spacing in the United States today is 30 inches.

John Deere MaxEmerge XP Planter with Case IH AFS
precision farming system which auto-steers using GPS.

Various machines meter out seeds for sowing in rows. The ones that handle larger seeds tend to be called planters, whereas the ones that handle smaller seeds tend to be called seed drills, grain drills, and seeders (including precision seeders). They all share a set of similar concepts in the ways that they work, but there is established usage in which the machines for sowing some crops including maize (corn), beans, and peas are mostly called planters, whereas those that sow cereals are drills.

On smaller and older planters, a marker extends out to the side half the width of the planter and creates a line in the field where the tractor should be centered for the next pass. The marker is usually a single disc harrow disc on a rod on each side of the planter. On larger and more modern planters, GPS navigation and auto-steer systems for the tractor are often used, eliminating the need for the marker. Some precision farming equipment such as Case IH AFS uses GPS/RKS and computer-controlled planter to sow seeds to precise position accurate within 2 cm. In an irregularly shaped field, the precision farming equipment will automatically hold the seed release over area already sewn when the tractor has to run overlapping pattern to avoid obstacles such as trees.

A Kinze 2200 planter.

Older planters commonly have a seed bin for each row and a fertilizer bin for two or more rows. In each seed bin plates are installed with a certain number of teeth and tooth spacing according to the type of seed to be sown and the rate at which the seeds are to be sown. The tooth size (actually the size of the space between the teeth) is just big enough to allow one seed in at a time but not big enough for two. Modern planters often have a large bin for seeds that are distributed to each row known as central commodity systems.

A class of planters that dig down farther than others are called listers. They are not used much anymore, as their use belonged to a set of high-till methods that low-till and no-till methods have largely replaced. Corn listers were common on the Great Plains in the 1920s through 1950s.

Drive Systems

There are different types of planters available with the main difference being mechanically driven vs. hydraulic/electrical driven. In a mechanical drive system the unit works by a small suspended tire being driven by another which is in contact with the ground (driven) tire. As the operator lowers the planter the two tires make contact and the planter is engaged. When the driven wheel begins to turn it then turns a series of gears that determine the population of the seed produced. The gears can be changed by the operator in order to change the planting population. A hydraulic driven system came about to correct the shortfalls of the ground driven system. Hydraulic driven systems allow the operator to change population on the go, as well as allowing the computer controller to follow a prepared prescription for an individual field. The system also allowed for plant populations to be infinite in that mechanical gears systems are limited to set number of population settings and gears available from manufactures. In 2014 John Deere introduced the ExactEmerge row unit which introduced high-speed planting. Precision Planting followed suit and released the vDrive system. These system were unique, not that they were electrical, but that they allowed an operator to double their speed when planting. Other manufacturers had already developed an electrical planter, but lacked these additional improvements. Traditionally, an operator would plant at about 4.5-5.5 mph for optimal performance. However, with the advent of these systems electrical motors match the speed of the tractor and "dead-drop" the seed in the trench using either a belt or brush-belt which cause the forward momentum of the planter to be offset by the rearward momentum of the seed. Older systems would instead drop the seed through a tube after the meter rather than place it in the seed trench directly.

Subsoiler

Howse brand modular subsoiler mounted to a tractor.

A subsoiler or flat lifter is a tractor-mounted farm implement used for deep tillage, loosening and breaking up soil at depths below the levels worked by moldboard ploughs, disc harrows, or rototillers. Most such tools will break up and turn over surface soil to a depth of 15–20 cm (5.9–7.9 in), whereas a subsoiler will break up and loosen soil to twice those depths. Typically a subsoiler mounted on a compact utility tractor will reach depths of about 30 cm (12 in) and typically have only one thin blade with a sharpened tip.

Modular subsoiler unit, unmounted with accessories.

The subsoiler is a tillage tool which will improve growth in all crops where soil compaction is a problem. In agriculture angled wings are used to lift and shatter the hardpan that builds up due to compaction. The design provides deep tillage, loosening soil depth is deeper than a tiller or plough is capable of reaching. Agricultural subsoilers, according to the Unverferth Company, can disrupt hardpan ground down to 60 cm (24 in) depths.

The subsoiler consists of three or more heavy vertical shanks (standards) mounted on a toolbar or frame with share bolts. They can be operated at depths of 45 to 75 cm or more. A ripper normally run 35-45cm deep. Shanks are curved and have replaceable tips. Each shank fitted with a replaceable point or foot, than the chisel plough to break through impervious layer shattering the sub-soil to a depth of 45 to 75 cm and requires 60 to 100 hp to pull a single subsoil point through a hard soil. Subsoiling is a slow operation and requires high power input. The shanks should be inclined to the vertical at an angle greater than 25-300, preferably 450, and it is advisable that the height be adjustable. The points of the shanks are normally about 30 cm wide and should be easy to replace. The conditions of the point is very important and often the subsoiler fails to give good results due to the conditions of its points. Point can be fitted with horizontal wings, about 300 mm wide, which considerably increases the width of soil below ploughing depth loosened by the subsoiler. These plows are sometimes equipped with a torpedo-shaped attachment for making subsurface drainage channels. The subsoilers are raised and lowered hydraulically. Some models feature power-take-off (PTO) driven vibrating devices. The typical spacing is 76 to 100 cm between shanks. Shanks should be able to reach 2.5 to 5 cm below the deepest compacted layer. Shank spacing and height should be adjustable in the field. Towed subsoilers should have gauge wheels to control the shank's depth.

Shanks usually are from 2 to 4 cm thick. Thinner shanks are suited for agricultural use. Thicker shanks hold up better in rocky conditions, but require larger, more powerful equipment to pull them and disturb the surface more.

Various manufacturers' brochures claim that crops perform well during hot and dry seasons because roots penetrate soil layers deeper to reach moisture and nutrients. Brochures further claim that in wet conditions, the water passes more easily through the shattered areas, reducing the possibility of crops drowning.

Agricultural subsoiler implements will have multiple deeper reaching blades; each blade is called a scarifier or shank. The common subsoilers for agricultural use are available with three, five or seven shanks. Subsoilers can be up to 15 feet (4.6 m) wide, some models are towed behind tractors while others are mounted to the three-point hitch.

One type of subsoiler has a torpedo-shaped tip and is called a mole plough because the tip describes a path much like the burrow that a mole creates. Mole ploughs are used to create tile drainage, with or without tiles or tile line added. A form of this implement (with a single blade), a pipe-and-cable-laying plough, is used to lay buried cables or pipes, without the need to dig a deep trench and re-fill it.

SEED DRILL

A seed drill machine which uses a shoe type coulter to place seeds underground.

A seed drill is a device that sows the seeds for crops by positioning them in the soil and burying them to a specific depth. This ensures that seeds will be distributed evenly.

The seed drill sows the seeds at the proper seeding rate and depth, ensuring that the seeds are covered by soil. This saves them from being eaten by birds and animals, or being dried up due to exposure to sun. With seed drill machines, seeds are distributed in rows, however the distance between seeds along the row cannot be adjusted by the user as in the case of vacuum precision planters. The distance between rows is typically set by the manufacturer. This allows plants to get sufficient sunlight, nutrients, and water from the soil. Before the introduction of the seed drill, most seeds were planted by hand broadcasting, an imprecise and wasteful process with a poor distribution of seeds and low productivity. Use of a seed drill can improve the ratio of crop yield (seeds harvested per seed planted) by as much as nine times. The use of seed drill saves time and labor.

Some machines for metering out seeds for planting are called planters. The concepts evolved from ancient Chinese practice and later evolved into mechanisms that pick up seeds from a bin and deposit them down a tube.

Seed drills of earlier centuries included single-tube seed drills in Sumer and multi-tube seed drills in China, and later a seed drill by Jethro Tull that was influential in the growth of farming technology in recent centuries. Even for a century after Tull, hand sowing of grain remained common.

Design

In older methods of planting, a field is initially prepared with a plow to a series of linear cuts known as furrows. The field is then seeded by throwing the seeds over the field, a method known as manual broadcasting. The seeds may not be sown to the right depth nor the proper distance from one another. Seeds that land in the furrows have better protection from the elements, and natural erosion or manual raking will cover them while leaving some exposed. The result is a field planted roughly in rows, but having a large number of plants outside the furrow lanes.

There are several downsides to this approach. The most obvious is that seeds that land outside the furrows will not have the growth shown by the plants sown in the furrow since they are too shallow on the soil. Because of this, they are lost to the elements. Many of the seeds remain on the surface where they are vulnerable to being eaten by birds or carried away on the wind. Surface seeds commonly never germinate at all or germinate prematurely, only to be killed by frost.

Since the furrows represent only a portion of the field's area, and broadcasting distributes seeds fairly evenly, this results in considerable wastage of seeds. Less obvious are the effects of over-seeding; all crops grow best at a certain density, which varies depending on the soil and weather conditions. Additional seeding above this will actually reduce crop yields, in spite of more plants being sown, as there will be competition among the plants for the minerals, water, and the soil available. Another reason is that the mineral resources of the soil will also deplete at a much faster rate, thereby directly affecting the growth of the plants.

The invention of the seed drill dramatically improved germination. The seed drill employed a series of runners spaced at the same distance as the plowed furrows. These runners, or drills, opened the furrow to a uniform depth before the seed was dropped. Behind the drills were a series of presses, metal discs which cut down the sides of the trench into which the seeds had been planted, covering them over.

This innovation permitted farmers to have precise control over the depth at which seeds were planted. This greater measure of control meant that fewer seeds germinated early or late and that seeds were able to take optimum advantage of available soil moisture in a prepared seedbed. The result was that farmers were able to use less seed and at the same time experience larger yields than under the broadcast methods.

Uses

Drilling is the term used for the mechanized sowing of an agricultural crop. Traditionally, a seed drill used to consist of a hopper filled with seeds arranged above a series of tubes that can be set at selected distances from each other to allow optimum growth of the resulting plants. Seeds are spaced out using fluted paddles which rotate using a geared drive from one of the drill's land wheels—seed rate is altered by changing gear ratios. Most modern drills use air to convey seed in plastic tubes from the seed hopper to the coulters—it is an arrangement which allows seed drills to be much wider than the seed hopper—as much as 12 m wide in some cases. The seed is metered mechanically into an air stream created by a hydraulically powered onboard fan and conveyed initially to a distribution head which sub-divides the seed into the pipes taking the seed to the individual colters.

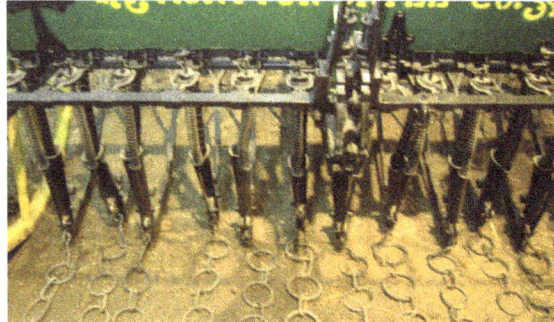

1902 model 12-run seed drill produced by Monitor Manufacturing Company, Minneapolis, Minnesota.

The seed drill allows farmers to sow seeds in well-spaced rows at specific depths at a specific seed rate; each tube creates a hole of a specific depth, drops in one or more seeds, and covers it over. This invention gives farmers much greater control over the depth that the seed is planted and the ability to cover the seeds without back-tracking. The result is an increased rate of germination, and a much-improved crop yield (up to eight times).

The use of a seed drill also facilitates weed control. Broadcast seeding results in a random array of growing crops, making it difficult to control weeds using any method other than hand weeding. A field planted using a seed drill is much more uniform, typically in rows, allowing weeding with the hoe during the growing season. Weeding by hand is laborious and inefficient. Poor weeding reduces crop yield, so this benefit is extremely significant.

Modern air seeder and hoe drill combination.

Before the operation of the seed drill, the ground must be plowed and harrowed. The plow digs up the earth and the harrow smooths the soil and breaks up any clumps. The drill must then be set for the size of the seed used. Afterwards, the grain is put in the hopper on top which then follows along behind the drill while it spaces and plants the seed. This system is still used today but has been modified and updated such that a farmer can plant many rows of seed at the same time.

A seed drill can be pulled across the field using bullocks or a tractor. Seeds sown using a seed drill are distributed evenly and placed at the correct depth in the soil.

HARROW

A spring-tooth drag harrow.

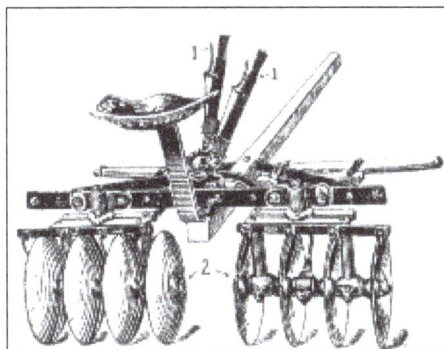
Disc harrows.

In agriculture, a harrow (often called a set of harrows in a plurale tantum sense) is an implement for breaking up and smoothing out the surface of the soil. In this way it is distinct in its effect from the plough, which is used for deeper tillage. Harrowing is often carried out on fields to follow the rough finish left by plowing operations. The purpose of this harrowing is generally to break up clods (lumps of soil) and to provide a finer finish, a good tilth or soil structure that is suitable for seedbed use. Coarser harrowing may also be used to remove weeds and to cover seed after sowing. Harrows differ from cultivators in that they disturb the whole surface of the soil, such as to prepare a seedbed, instead of disturbing only narrow trails that skirt crop rows (to kill weeds).

Crumbler roller, commonly used to compact soil after it has been loosened by a harrow.

Clydesdale horses pulling spike harrows, Murrurundi, New South Wales, Australia.

There are four general types of harrows: disc harrows, tine harrows (including spring-tooth harrows, drag harrows, and spike harrows), chain harrows, and chain-disk harrows. Harrows were originally drawn by draft animals, such as horses, mules, or oxen, or in some times and places by manual labourers. In modern practice they are almost always tractor-mounted implements, either trailed after the tractor by a drawbar or mounted on the three-point hitch.

A modern development of the traditional harrow is the rotary power harrow, often just called a power harrow.

Types

In cooler climates the most common types are the disc harrow, the chain harrow, the tine harrow or spike harrow and the spring tine harrow. Chain harrows are often used for lighter work such as levelling the tilth or covering seed, while disc harrows are typically used for heavy work, such as following ploughing to break up the sod. In addition, there are various types of power harrow, in which the cultivators are power-driven from the tractor rather than depending on its forward motion.

Tine harrows are used to refine seed-bed condition before planting, to remove small weeds in growing crops and to loosen the inter-row soils to allow for water to soak into the subsoil. The fourth is a chain disk harrow. Disk attached to chains are pulled at an angle over the ground. These harrows move rapidly across the surface. The chain and disk rotate to stay clean while breaking up the top surface to about 1 inch (3 cm) deep. A smooth seedbed is prepared for planting with one pass.

Harrowing with tractor and disk harrow in the 1940s.

Chain harrowing can be used on pasture land to spread out dung, and to break up dead material (*thatch*) in the sward, and similarly in sports-ground maintenance a light chain harrowing is often used to level off the ground after heavy use, to remove and smooth out boot marks and indentations. Used on tilled land in combination with the other two types, chain harrowing rolls remaining larger soil clumps to the surface where weather breaks them down and prevents interference with seed germination.

All four harrow types can be used in one pass to prepare soil for seeding. It is also common to use any combination of two harrows for a variety of tilling processes. Where harrowing provides a very fine tilth, or the soil is very light so that it might easily be wind-blown, a roller is often added as the last of the set.

Harrows may be of several types and weights, depending on their purpose. They almost always consist of a rigid frame that holds discs, teeth, linked chains, or other means of moving soil—but tine and chain harrows are often only supported by a rigid towing-bar at the front of the set.

In the southern hemisphere, so-called *giant discs* are a specialised kind of disc harrows that can stand in for a plough in rough country where a mouldboard plough cannot handle tree-stumps and rocks, and a disc-plough is too slow (because of its limited number of discs). Giant scalloped-edged discs operate in a set, or frame, that is often weighted with concrete or steel blocks to improve penetration of the cutting edges. This sort of cultivation is usually followed by broadcast fertilisation and seeding, rather than drilled or row seeding. A drag is a heavy harrow.

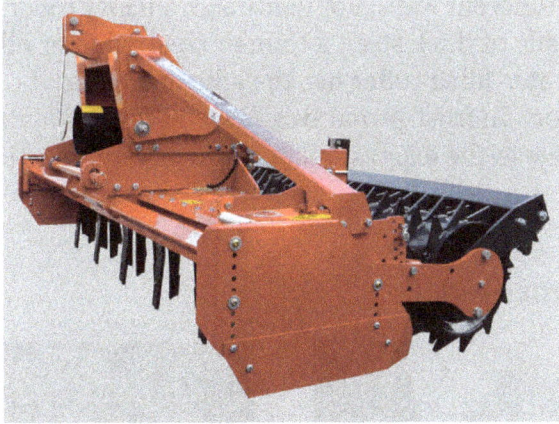

Power Harrow.

Power Harrow

A rotary power harrow, or simply power harrow, has multiple sets of vertical tines. Each set of tines is rotated on a vertical axis and tills the soil horizontally. The result is that, unlike a rotary tiller, soil layers are not turned over or inverted, which is useful in preventing dormant weed seeds from being brought to the surface, and there is no horizontal slicing of the subsurface soil that can lead to hardpan formation.

ROLLER

The roller is an agricultural tool used for flattening land or breaking up large clumps of soil, especially after ploughing or disc harrowing. Typically, rollers are pulled by tractors or, prior to mechanisation, a team of animals such as horses or oxen. As well as for agricultural purposes, rollers are used on cricket pitches and residential lawn areas.

A roller in a typical power farming application.

Flatter land makes subsequent weed control and harvesting easier, and rolling can help to reduce moisture loss from cultivated soil. On lawns, rolling levels the land for mowing and compacts the soil surface.

Rollers may be weighted in different ways. For many uses a heavy roller is used. These may consist of one or more cylinders made of thick steel, a thinner steel cylinder filled with concrete, or a cylinder filled with water. A water-filled roller has the advantage that the water may be drained out for lighter use or for transport. In frost-prone areas a water filled roller must be drained for winter storage to avoid breakage due to the expansion for water as it turns to ice.

Designs

One-piece versus Segmented

A 12-foot smooth roller comprising eight 1.5-foot segments.

A field after rolling with a Cambridge (or similar) roller.

On tilled soil a one-piece roller has the disadvantage that when turning corners the outer end of the roller has to rotate much faster than the inner end, forcing one or both ends to skid. A one-piece roller turned on soft ground will skid up a heap of soil at the outer radius, leaving heaps, which is counter-productive. Rollers are often made in two or three sections to reduce this problem, and the Cambridge roller overcomes it altogether by mounting many small segments onto one axle so that they can each rotate at local ground-speed.

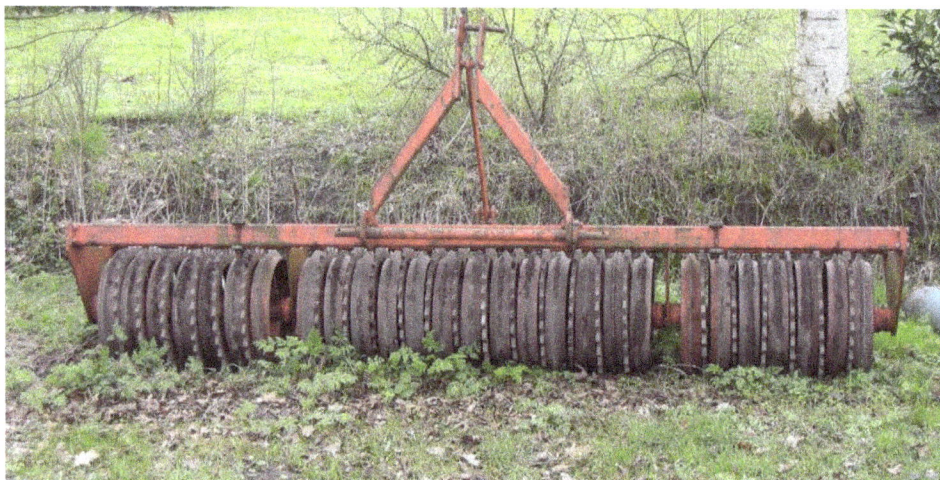

A ridged roller comprising many segments is usually called a Cambridge roller in the United Kingdom and a cultipacker in the United States; each name originated with a manufacturer in the respective country and evolved into the regionally prevalent name for the type.

Smooth versus Ridged

The surface of rollers may be smooth, or it may be textured to help break up soil or to groove the final surface to reduce scouring from rain. Each segment of a Cambridge roller has a rib around its edge for this purpose. The name cultipacker is often used for such ridged types, especially in the United States.

Uses

Farming

Rollers are a secondary tillage tool used for flattening land or breaking up large clumps of soil, especially after ploughing or disc harrowing. Rollers are typically pulled by tractors today. Before mechanised agriculture, a team of working animals such as horses or oxen provided the power. Animal power is still used today in some contexts, such as on Amish farms in the United States and in regions of Asia where draft oxen are still widely used.

Rollers prepare optimal seedbeds by making them as flat as is practical and moderately firmed. Flatness is important at planting because it is the only practical way to control average seed planting depth without laborious hand planting of each seed; it is not practical to follow an instruction of 1-cm planting depth if the contour of the seedbed varies by 2 cm or more between adjacent spots. This is why breaking up of even small clods/lumps, and well-leveled spreading of soil, is important at planting time.

Flatter land also makes subsequent weed control and harvesting easier. For example, in mechanical weed control, controlling cultivator tooth depth is practical only with a decently flat soil contour, and in combining, controlling combine head height is practical only with a decently flat soil contour. Rolling is also believed to help reduce moisture loss from cultivated soil.

Ganging and Trailing

Rollers may be ganged to increase the width of each pass/swath. Rollers may be trailed after other equipment such as ploughs, disc harrows, or mowers.

Cricket Pitch

In cricket, rollers are used to make the pitch flat and less dangerous for batsmen. Several size rollers have been used in the history of cricket, from light rollers that were used in the days of uncovered pitches and at some stages during the 1950s to make batting less easy, to the modern "heavy roller" universally used in top-class cricket today. Regulations permit a pitch only to be rolled at the commencement of each innings or day's play, but this has still had a massive influence on the game by eliminating the shooters that were ubiquitous on all but light soils before heavy rollers were used. Heavy rollers have sometimes been criticised for making batting too easy and for reducing the rate at which pitches dry out after rain in the cool English climate.

Lawn

Lawn rollers are designed to even out or firm up the lawn surface, especially in climates where heaving causes the lawn to be lumpy. Heaving may result when the ground freezes and thaws many times over winter. Where this occurs, gardeners are advised to give the lawn a light rolling with a lawn roller in the spring. Clay or wet soils should not be rolled as they become compacted.

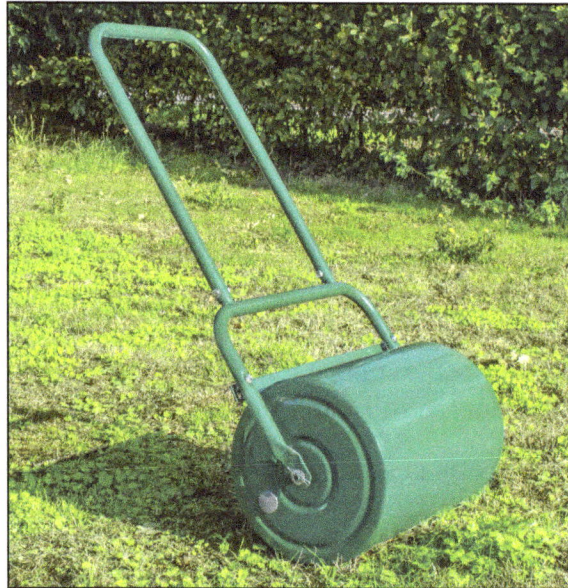

Lawn roller.

CULTIVATOR

Cultivator is a farm implement or machine designed to stir the soil around a crop as it matures to promote growth and destroy weeds.

Horse-drawn cultivators were introduced in the mid-19th century. By 1870 a farmer with two horses could cultivate as much as 15 acres (6 hectares) a day with a machine the shovels (blades) of which straddled the crop rows. In the 20th century, with tractor power substituted for horses, the number of rows a single machine could cultivate grew to equal the capacity of multiple-row planters. Typical shovels are short, narrow, slightly curved, pointed steel pieces with polished front surfaces that dig into the soil in proportion to the pressure applied. The kind and number used per gang (in a single mounting) depend on crop and soil characteristics. Rotary hoes, used for early cultivation of corn, cotton, soybeans, potatoes, and small grain, have as many as 12 sections, each mounting several hoe wheels, with the whole machine up to 40 feet (12 metres) wide. When the rotary hoe is drawn backward, it gives a treading action that crushes clods and pulverizes stalks.

Spring-tooth weeders have light spring teeth that flick out shallow-rooted weeds without injuring growing plants and can therefore be operated directly over planted rows in an early stage, ridding the field of many weeds as they emerge. Rod weeders are used for weed control in open unplanted fields; their working element is a square-section rod that revolves a few inches below the soil surface. Field cultivators, essentially light plows, are equipped with spring teeth, shovels, or sweeps.

MOWER

A mower is machine that cuts (mows) grass or other plants that grow on the ground. Usually mowing is distinguished from reaping, which uses similar implements, but is the traditional term for harvesting grain crops, e.g. with reapers and combines.

A smaller mower used for lawns and sports grounds (playing fields) is called a lawn mower or grounds mower, which is often self-powered, or may also be small enough to be pushed by the operator. Grounds mowers have reel or rotary cutters. Larger mowers or mower-conditioners are mainly used to cut grass (or other crops) for hay or silage and often place the cut material into rows, which are referred to as windrows. Swathers (or windrowers) are also used to cut grass (and grain crops). Prior to the invention and adoption of mechanized mowers, (and today in places where use a mower is impractical or uneconomical), grass and grain crops were cut by hand using scythes or sickles.

Mower Configurations

Larger mowers are usually *ganged* (equipped with a number or gang of similar cutting units), so they can adapt individually to ground contours. They may be powered and drawn by a tractor or draft animals. The cutting units can be mounted underneath the tractor between the front and rear wheels, mounted on the back with a three-point hitch or pulled behind the tractor as a trailer. There are also dedicated self-propelled cutting machines, which often have the mower units mounted at the front and sides for easy visibility by the driver. *Boom* or *side-arm* mowers are mounted on long hydraulic arms, similar to a backhoe arm, which allows the tractor to mow steep banks or around objects while remaining on a safer surface.

Mower Types

The cutting mechanism in a mower may be one of several different designs:

Sickle Mower

Eicher tractor with a mid-mounted finger-bar mower.

Sickle mowers, also called reciprocating mowers, bar mowers, sickle-bar mowers, or finger-bar mowers, have a long (typically six to seven and a half feet) bar on which are mounted fingers with stationary guardplates. In a channel on the bar there is a reciprocating sickle with very sharp sickle sections (triangular blades). The sickle bar is driven back and forth along the channel. The grass, or other plant matter, is cut between the sharp edges of the sickle sections and the finger-plates (this action can be likened to an electric hair clipper).

The bar rides on the ground, supported on a skid at the inner end, and it can be tilted to adjust the height of the cut. A spring-loaded board at the outer end of the bar guides the cut hay away from the uncut hay. The so-formed channel, between cut and uncut material, allows the mower skid to ride in the channel and cut only uncut grass cleanly on the next swath. These were the first successful horse-drawn mowers on farms and the general principles still guide the design of modern mowers.

Rotary Mower

Rotary cutters mounted on a swather.

Rotary mowers, also called *drum mowers*, have a rapidly rotating bar, or disks mounted on a bar, with sharpened edges that cut the crop. When these mowers are tractor-mounted they are easily capable of mowing grass at up to 20 miles per hour (32 km/h) in good conditions. Some models are designed to be mounted in double and triple sets on a tractor, one in the front and one at each side, thus able to cut up to 20 foot (6 metre) swaths.

In rough cutting conditions, the blades attached to the disks are swivelled to absorb blows from obstructions. Mostly these are rear-mounted units and in some countries are called *scrub cutters*. Self-powered mowers of this type are used for rougher grass in gardening and other land maintenance.

Reel Mower

Reel mowers, also called cylinder mowers (familiar as the hand-pushed or self-powered cylinder lawn mower), have a horizontally rotating cylindrical reel composed of helical blades, each of which in turn runs past a horizontal cutter-bar, producing a continuous scissor action. The bar is held at an adjustable level just above the ground and the reel runs at a speed dependent on the forward movement speed of the machine, driven by wheels running on the ground (or in self-powered applications by a motor). The cut grass may be gathered in a collection bin.

This type of mower is used to produce consistently short and even grass on bowling greens, lawns, parks and sports grounds. When pulled by a tractor (or formerly by a horse), these mowers are often ganged into sets of three, five or more, to form a *gang mower*. A well-designed reel mower can cut quite tangled and thick tall grass, but this type works best on fairly short, upright vegetation, as taller vegetation tends to be rolled flat rather than cut.

Reel mower

Home reel mowers have certain benefits over motor-powered mowers as they are quieter and not dependent on any extra form of power besides the person doing the mowing. This is useful not only to lessen dependence on other types of power which may have availability issues, but also lessens the impact on the environment.

Flail Mower

Flail mower

Flail mowers have a number of small blades on the end of chains attached to a horizontal axis. The cutting is carried out by the ax-like heads striking the grass at speed. These types are used on rough ground, where the blades may frequently be fouled by other objects, or on tougher vegetation than grass, such as brush (scrub). Due to the length of the chains and the higher weight of the blades, they are better at cutting thick brush than other mowers, because of the relatively high inertia of the blades. In some types the cut material may be gathered in a collection bin. As a boom mower, a flail mower may be used in an upright position for trimming the sides of hedges, when it is often called a hedge-cutter.

Drum Mower

Drum mowers have their horizontally-mounted cutting blades attached to the outside of a relatively large diameter disc fixed to the bottom of a smaller diameter drum and are principally designed for cutting lighter crops, such as grass, very quickly. The drive mechanism is top-mounted and often in the form of fully enclosed, bevel geared drive shafts.

Radon Drum Mower.

BALER

A Claas large round baler.

A baler, most often called a hay baler is a piece of farm machinery used to compress a cut and raked crop (such as hay, cotton, flax straw, salt marsh hay, or silage) into compact bales that are easy to handle, transport, and store. Often, bales are configured to dry and preserve some intrinsic (e.g. the nutritional) value of the plants bundled. Different types of balers are commonly used, each producing a different type of bale – rectangular or cylindrical, of various sizes, bound with twine, strapping, netting, or wire.

Industrial balers are also used in material recycling facilities, primarily for baling metal, plastic, or paper for transport.

Before the 19th century, hay was cut by hand and most typically stored in haystacks using hay forks to rake and gather the scythed grasses into optimal sized heaps — neither too large (promoting conditions that might create spontaneous combustion), nor too small, so much of the pile is susceptible to rotting. These haystacks lifted most of the plant fibers up off the ground, letting air in and water drain out, so the grasses could dry and cure, to retain nutrition for livestock feed at a later time. In the 1860s, mechanical cutting devices were developed; from these came the modern devices including mechanical mowers and balers. In 1872, a reaper that used a knotter device to bundle and bind hay was invented by Charles Withington; this was commercialized in 1874 by Cyrus McCormick. In 1936, Innes invented an automatic baler that tied bales with twine using Appleby-type knotters from a John Deere grain binder. In 1938 Edwin Nolt filed a patent for an improved version that was more reliable. The first round baler was probably invented in the late 19th century and one was shown in Paris by Pilter. This was a portable machine designed for use with threshing machines.

Round Baler

Allis Chalmers Rotobaler.

The most common type of baler in industrialized countries today is the round baler. It produces cylinder-shaped "round" or "rolled" bales. The design has a "thatched roof" effect that withstands weather well. Grass is rolled up inside the baler using rubberized belts, fixed rollers, or a combination of the two. When the bale reaches a predetermined size, either netting or twine is wrapped around it to hold its shape. The back of the baler swings open, and the bale is discharged. The bales are complete at this stage, but they may also be wrapped in plastic sheeting by a bale wrapper, either to keep hay dry when stored outside or convert damp grass into silage. Variable-chamber large round balers typically produce bales from 48 to 72 inches (120 to 180 cm) in diameter and up to 60 inches (150 cm) in width. The bales can weigh anywhere from 1,100 to 2,200 pounds (500 to 1,000 kg), depending upon size, material, and moisture content. Common modern small round balers (also called "mini round balers" or "roto-balers") produce bales 20 to 22 inches (51 to 56 cm) in diameter and 20.5 to 28 inches (52 to 71 cm) in width, generally weighing from 40 to 55 pounds (18 to 25 kg).

Originally conceived by Ummo Luebben circa 1910, the first round baler did not see production until 1947 when Allis-Chalmers introduced the Roto-Baler. Marketed for the water-shedding and light weight properties of its hay bales, AC had sold nearly 70,000 units by the end of production in 1960. The next major innovation began in 1965 when a graduate student at Iowa State University, Virgil Haverdink, sought out Wesley F. Buchele, a professor of Agricultural Engineering, seeking a research topic for a master thesis. Over the next year Buchele and Haverdink developed a new design for a large round baler, completed and tested in 1966, and thereafter dubbed the Buchele-Haverdink large round baler. The large round bales were about 1.5 meters (4.9 feet) in diameter, 2 meters (6.6 feet) long, and they weighed about 270 kilograms (600 pounds) after they dried—about 80 kg/m³ (5 lb/ft³). The design was promoted as a "Whale of a Bale" and Iowa State University now explains the innovative design as follows:

"Farmers were saved from the backbreaking chore of slinging hay bales in the 1960s, when Iowa State agricultural engineering professor Wesley Buchele and a group of student researchers invented a baler that produced large, round bales that could be moved by tractor. The baler has become the predominant forage-handling machine in the United States."

In the summer of 1969, the Australian Econ Fodder Roller baler came out, a design that made a 135 kg (298 lb) ground-rolled bale. In September of that same year, The Hawkbilt Company of Vinton, Iowa, contacted Dr. Buchele about his design, then fabricated a large ground-rolling round baler which baled hay that had been laid out in a windrow, and began manufacturing large round balers in 1970. In 1972, Gary Vermeer of Pella, Iowa, designed and fabricated a round baler after the design of the A-C Roto-Baler, and the Vermeer Company began selling its model 605 - the first modern round baler. The Vermeer design used belts to compact hay into a cylindrical shape as is seen today. In the early 1980s, collaboration between Walterscheid and Vermeer produced the first effective uses of CV joints in balers, and later in other farm machinery. Due to the heavy torque required for such equipment, double Cardan joints are primarily used. Former Walterscheid engineer Martin Brown is credited with "inventing" this use for universal joints. By 1975, fifteen American and Canadian companies were manufacturing large round balers.

Transport, Handling and Feeding

Short-haul Transport and On-field Handling

A large round bale.

Due to the ability for round bales to roll away on a slope, they require specific treatment for safe transport and handling. Small round bales can typically be moved by hand or with lower-powered equipment. Large round bales, due to their size and weight (they can weigh a ton or more) require special transport and moving equipment.

The most important tool for large round bale handling is the bale spear or spike, which is usually mounted on the back of a tractor or the front of a skid-steer. It is inserted into the approximate center of the round bale, then lifted and the bale is hauled away. Once at the destination, the bale is set down, and the spear pulled out. Careful placement of the spear in the center is needed or the bale can spin around and touch the ground while in transport, causing a loss of control. When used for wrapped bales that are to be stored further, the spear makes a hole in the wrapping that must be sealed with plastic tape to maintain a hermetic seal.

Alternatively, a grapple fork may be used to lift and transport large round bales. The grapple fork is a hydraulically driven implement attached to the end of a tractor's bucket loader. When the hydraulic cylinder is extended, the fork clamps downward toward the bucket, much like a closing hand. To move a large round bale, the tractor approaches the bale from the side and places the bucket underneath the bale. The fork is then clamped down across the top of the bale, and the bucket is lifted with the bale in tow. Grab hooks installed on the bucket of a tractor are another tool used to handle round bales, and be done by a farmer with welding skills by welding two hooks and a heavy chain to the outside top of a tractor front loader bucket.

Long-haul Transport

The rounded surface of round bales poses a challenge for long-haul, flat-bed transport, as they could roll off of the flat surface if not properly supported. This is particularly the case with large round bales; their size makes them difficult to flip, so it may not be feasible to flip many of them onto the flat surface for transport and then re-position them on the round surface at the destination. One option that works with both large and small round bales is to equip the flat-bed trailer with guard-rails at either end, which prevent bales from rolling either forward or backward. Another solution is the saddle wagon, which has closely spaced rounded saddles or support posts in which round bales sit. The tall sides of each saddle prevent the bales from rolling around while on the wagon, as the bale settles down in between posts. On 3 September 2010, on the A381 in Halwell near Totnes, Devon, UK an early member of British rock group ELO Mike Edwards was killed when his van was crushed by a large round bale. The cellist, 62, died instantly when the 600-kilogram (1,300 lb) bale fell from a tractor on nearby farmland before rolling onto the road and crushing his van.

Feeding

A large round bale can be directly used for feeding animals by placing it in a feeding area, tipping it over, removing the bale wrap, and placing a protective ring (a *ring feeder*) around the outside so that animals don't walk on hay that has been peeled off the outer perimeter of the bale. The round baler's rotational forming and compaction process also enables both large and small round bales to be fed out by unrolling the bale, leaving a continuous flat strip in the field or behind a feeding barrier.

Silage

Silage, a fermented animal feed, was introduced in the late 1800s and can also be stored in a silage or haylage bale, which is a high-moisture bale wrapped in plastic film. These are baled much wetter than hay bales, and are usually smaller than hay bales because the greater moisture content makes them heavier and harder to handle. These bales begin to ferment almost immediately, and the metal bale spear stabbed into the core becomes very warm to the touch from the fermentation process.

Silage or haylage bales may be wrapped by placing them on a rotating bale spear mounted on the rear of a tractor. As the bale spins, a layer of plastic cling film is applied to the exterior of the bale.

This roll of plastic is mounted in a sliding shuttle on a steel arm and can move parallel to the bale axis, so the operator does not need to hold up the heavy roll of plastic. The plastic layer extends over the ends of the bale to form a ring of plastic approximately 12 inches (30 cm) wide on the ends, with hay exposed in the center.

To stretch the cling-wrap plastic tightly over the bale, the tension is actively adjusted with a knob on the end of the roll, which squeezes the ends of the roll in the shuttle. The operator recovers by quickly loosening the tension and allows the plastic to feed out halfway around the bale before reapplying the tension to the sheeting.

These bales are placed in a long continuous row, with each wrapped bale pressed firmly against all the other bales in the row before being set down onto the ground. The plastic wrap on the ends of each bale sticks together to seal out air and moisture, protecting the silage from the elements. The end-bales are hand-sealed with strips of cling plastic across the opening.

The airtight seal between each bale permits the row of round bales to ferment as if they were in a silo bag, but they are easier to handle than a silo bag, as they are more robust and compact. The plastic usage is relatively high, and there is no way to reuse the silage-contaminated plastic sheeting, although it can be recycled or used as a fuel source via incineration. The wrapping cost is approximately US $5 per bale.

An alternative form of wrapping uses the same type of bale placed on a bale wrapper, consisting of pair of rollers on a turntable mounted on the three-point linkage of a tractor. It is then spun about two axes while being wrapped in several layers of cling-wrap plastic film. This covers the ends and sides of the bale in one operation, thus sealing it separately from other bales. The bales are then moved or stacked using a special pincer attachment on the front loader of a tractor, which does not damage the film seal. They can also be moved using a standard bale spike, but this punctures the airtight seal, and the hole in the film must be repaired after each move.

Plastic-wrapped bales must be unwrapped before being fed to livestock to prevent accidental ingestion of the plastic. Like round hay bales, silage bales are usually fed using a *ring feeder*.

Large rectangular baler.

Large Rectangular Baler

In 1978, Hesston introduced the first "large square baler," capable of compacting hay into more easily transported large square bales that could be stacked and tarped in the field (to protect them from rain) or loaded on trucks or containers for trucking or export. Depending upon the baler, these bales can weigh from 1000 pounds to 2200 pounds for a 3'x3'x9' or 3'x4'x9' bale (versus 900 pounds for a 3'x4' round bale). As the pickup revolves just above the ground surface, the tines pick up and feed the hay into the flake forming chamber, where a "flake" of hay is formed before being pushed up into the path of the plunger, which then compresses it with great force (200 to over 750 kilonewtons, depending on model) against the existing bale in the chamber. Once the desired length is achieved, the knotter arm is mechanically tripped to begin the knotting cycle in which several knotters (4-6 is common) tie the 4-6 strings that maintain the bale's shape. In the prairies of Canada, the large rectangular balers are also called "prairie raptors".

Rectangular Bale Handling and Transport

Large rectangular bales in a field, Charente, France. Sizes of stacks of baled hay need to be carefully managed, as the curing process is exothermic and the built up heat around internal bales can reach ignition temperatures in the right weather history and atmospheric conditions. Building a deep stack either too wide, or too high increases the risk of spontaneous ignition.

Rectangular bales are easier to transport than round bales, since there is little risk of the bale rolling off the back of a flatbed trailer. The rectangular shape also saves space and allows a complete solid slab of hay to be stacked for transport and storage. Most balers allow adjustment of length and it is common to produce bales of twice the width, allowing stacks with brick-like alternating groups overlapping the row below at right angles, creating a strong structure.

They are well-suited for large-scale livestock feedlot operations, where many tons of feed are rationed every hour. Most often, they are baled small enough that one person can carry or toss them where needed.

Due to the huge rectangular shape, large spear forks, or squeeze grips are mounted to heavy lifting machinery, such as large fork lifts, tractors equipped with front end loaders, telehandlers, hay squeezes or wheel loaders, to lift these bales.

Small Rectangular Baler

A type of baler that produces small rectangular (often called "square") bales was once the most prevalent form of baler, but is less common today. It is primarily used on small acreages where large equipment is impractical, and also for the production of hay for small operations, particularly horse owners who may not have access to the specialized feeding machinery used for larger bales. Each bale is about 15 by 18 by 40 inches (38 cm × 46 cm × 102 cm). The bales are usually wrapped with two, but sometimes three, or more strands of knotted twine. The bales are light enough for one person to handle, about 45 to 60 pounds (20 to 27 kg), depending upon the crop and pressure applied (can be 100 lbs for a 16" × 18" 2-string bale). Many balers have adjustable bale chamber pressure and bale length, so shorter, less-dense bales can be produced for ease of handling.

A small square baler.

To form the bale, the material to be baled (which is often hay or straw) in the windrow is lifted by tines in the baler's reel. This material is then packed into the bale chamber, which runs the length of one side of the baler (normally the right hand side when viewed from the front) in offset balers. Balers like Hesston models use an in-line system where the hay goes straight through from the pickup to the flake chamber to the plunger and bale-forming chamber. A combination plunger and knife move back and forth in the front of this chamber, with the knife closing the door into the bale chamber as it moves backwards. The plunger and knife are attached to a heavy asymmetrical fly-wheel to provide extra force as they pack the bales. A measuring device—normally a spiked wheel that is turned by the emerging bales—measures the amount of material that is being compressed and, at the appropriate length it triggers the knotters that wrap the twine around the bale and tie it off. As the next bale is formed the tied one is driven out of the rear of the baling chamber, where it can either drop to the ground, or sent to a wagon towed behind the baler. When a wagon is used, the bale may be lifted by hand from the chamber by a worker on the wagon who stacks the bales on the wagon, or the bale may be propelled into the wagon by a mechanism on the baler, commonly either a "thrower" (parallel high-speed drive belts which throw the bale into the wagon) or a "kicker" (mechanical arm which throws the bale into the wagon). In the case of a thrower or kicker, the wagon has high walls on the left, right, and back sides, and a short wall on the front side, to contain the randomly piled bales. This process continues as long as there is material to be baled, and twine to tie it with.

This form of bale is not much used in large-scale commercial agriculture, because of the costs involved in handling many small bales. However, it enjoys some popularity in small-scale, low-mechanization agriculture and horse-keeping. Besides using simpler machinery and being easy to handle, these small bales can also be used for insulation and building materials in straw-bale construction. Square bales may generally weather better than round bales because a more much dense stack can be put up. However, they don't shed water as round bales do. Convenience is also a major factor in farmers deciding to continue putting up square bales, as they make feeding and bedding in confined areas (stables, barns, etc.) much easier.

Many of these older balers are still to be found on farms today, particularly in dry areas, where bales can be left outside for long periods.

The automatic-baler for small square bales took on most of its present form in 1938 with the first such baler sold as Arthur S. Young's Automation Baler. It was manufactured in small numbers until acquired by New Holland Agriculture.

In Europe, in as early as 1939, both Claas of Germany and Rousseau SA of France had automatic twine tying pick-up balers. Most of these produced low density bales though. The first successful pick-up balers were made by the Ann Arbor Company in 1929. Ann Arbor was acquired by the Oliver Farm Equipment Company in 1943. Despite their head start on the rest of the field, no Ann Arbor balers carried automatic knotters or twisters and Oliver didn't produce its own automatic tying baler until 1949.

Hay Presses and Wire Balers

Stationary baler.

Prior to 1937 the hay press was the common name of the stationary baling implement, powered with a tractor or stationary engine using a belt on a belt pulley, with the hay being brought to the baler and fed in by hand. Later, balers were made mobile, with a 'pickup' to gather up the hay and feed it into the chamber. These often used air cooled gasoline engines mounted on the baler for power. The biggest change to this type of baler since 1940 is being powered by the tractor through its power take-off (PTO), instead of by a built-in internal combustion engine. In present-day production, small square balers can be ordered with twine knotters or wire tie knotters.

Not all stationary wire tying balers used 2 wires. It was not uncommon for the larger bale size (usually 17" × 22") machines to use 'boards' that had three slots for wires and hence tied three wires per bale. Most North American manufacturers produced these machines as either regular models or as size options. 'Small square' three wire tying pick-up balers were available from the early 1930s, principally from J. I. Case & Co. and Ann Arbor. These machines were hand tying and hand threading machines.

Pickup and Handling Methods

In the 1940s most farmers would bale hay in the field with a small tractor with 20 or less horsepower, and the tied bales would be dropped onto the ground as the baler moved through the field. Another team of workers with horses and a flatbed wagon would come by and use a sharp metal hook to grab the bale and throw it up onto the wagon while an assistant stacks the bale, for transport to the barn.

A later time-saving innovation was to tow the flatbed wagon directly behind the baler, and the bale would be pushed up a ramp to a waiting attendant on the wagon. The attendant hooks the bale off the ramp and stacks it on the wagon, while waiting for the next bale to be produced.

Eventually, as tractor horsepower increased, the thrower-baler became possible, which eliminated the need for someone to stand on the wagon and pick up the finished bales. The first thrower mechanism used two fast-moving friction belts to grab finished bales and throw them at an angle up in the air onto the bale wagon. The bale wagon was modified from a flatbed into a three-sided skeleton frame open at the front, to act as a catcher's net for the thrown bales.

As tractor horsepower further increased, the next innovation of the thrower-baler was the hydraulic tossing baler. This employs a flat pan behind the bale knotter. As bales advance out the back of the baler, they are pushed onto the pan one at a time. When the bale has moved fully onto the pan, the pan suddenly pops up, pushed by a large hydraulic cylinder, and tosses the bale up into the wagon like a catapult.

The pan-thrower method puts much less stress on the bales compared to the belt-thrower. The friction belts of the belt-thrower stress the twine and knots as they grip the bale, and would occasionally cause bales to break apart in the thrower or when the bales landed in the wagon.

Bales may be picked up from the field and stacked using a self-powered machine called a bale stacker, bale wagon or harobed. There are several designs and sizes. One type picks up square bales are dropped by the baler with the strings facing upward. The stacker will drive up to each bale, pick it up and set it on a three-bale-wide table (the strings are now facing upwards). Once three bales are on the table, the table lifts up and back, causing the three bales to face strings to the side again; this happens three more times until there are 16 bales on the main table. This table will lift like the smaller one, and the bales will be up against a vertical table. The machine will hold 160 bales (ten tiers); usually there will be cross-tiers near the center to keep the stack from swaying or collapsing if any weight is applied to the top of the stack. The full load will be transported to a barn; the whole rear of the stacker will tilt upwards until it is vertical. There will be two pushers that will extend through the machine and hold the bottom of the stack from being pulled out from the stacker while it is driven out of the barn.

Square bale stacker.

In Britain (if small square bales are still to be used), they are usually collected as they fall out of the baler in a *bale sledge* dragged behind the baler. This has four channels, controlled by automatic mechanical balances, catches and springs, which sort each bale into its place in a square *eight*. When the sledge is full, a catch is tripped automatically, and a door at the rear opens to leave the eight lying neatly together on the ground. These may be picked up individually and loaded by hand, or they may be picked up all eight together by a *bale grab* on a tractor, a special front loader consisting of many hydraulically powered downward-pointing curved spikes. The square eight will then be stacked, either on a trailer for transport, or in a roughly cubic field stack eight or ten layers high. This cube may then be transported by a large machine attached to the three-point hitch behind a tractor, which clamps the sides of the cube and lifts it bodily.

A smaller type of stacker.

Storage Methods

Before electrification occurred in rural parts of the United States in the 1940s, some small dairy farms would have tractors but not electric power. Often just one neighbor who could afford a tractor would do all the baling for surrounding farmers still using horses.

To get the bales up into the hayloft, a pulley system ran on a track along the peak of the barn's hayloft. This track also stuck a few feet out the end of the loft, with a large access door under the track. On the bottom of the pulley system was a bale spear, which is pointed on the end and has retractable retention spikes.

A flatbed wagon would pull up next to the barn underneath the end of the track, the spear lowered down to the wagon, and speared into a single bale. The pulley rope would be used to manually lift the bale up until it could enter the mow through the door, then moved along the track into the barn and finally released for manual stacking in tight rows across the floor of the loft. As the stack filled the loft, the bales would be lifted higher and higher with the pulleys until the hay was stacked all the way up to the peak.

When electricity arrived, the bale spear, pulley and track system were replaced by long motorized bale conveyors known as hay elevators. A typical elevator is an open skeletal frame, with a chain that has dull 3-inch (76 mm) spikes every few feet along the chain to grab bales and drag them along. One elevator replaced the spear track and ran the entire length of the peak of the barn. A second elevator was either installed at a 30-degree slope on the side of the barn to lift bales up to the peak elevator, or used dual front-back chains surrounding the bale to lift bales straight up the side of the barn to the peak elevator.

A bale wagon pulled up next to the lifting elevator, and a farm worker placed bales one at a time onto the angled track. Once bales arrived at the peak elevator, adjustable tipping gates along the length of the peak elevator were opened by pulling a cable from the floor of the hayloft, so that bales tipped off the elevator and dropped down to the floor in different areas of the loft. This permitted a single elevator to transport hay to one part of a loft and straw to another part.

This complete hay elevator lifting, transport, and dropping system reduced bale storage labor to a single person, who simply pulls up with a wagon, turns on the elevators and starts placing bales on it, occasionally checking to make sure that bales are falling in the right locations in the loft.

The neat stacking of bales in the loft is often sacrificed for the speed of just letting them fall and roll down the growing pile in the loft, and changing the elevator gates to fill in open areas around the loose pile. But if desired, the loose bale pile dropped by the elevator could be rearranged into orderly rows between wagon loads.

Usage Once in the Barn

The process of retrieving bales from a hayloft has stayed relatively unchanged from the beginning of baling. Typically workers were sent up into the loft, to climb up onto the bale stack, pull bales off the stack, and throw or roll them down the stack to the open floor of the loft. Once the bale is down on the floor, workers climb down the stack, open a cover over a bale chute in the floor of the loft, and push the bales down the chute to the livestock area of the barn.

Most barns were equipped with several chutes along the sides and in the center of the loft floor. This permitted bales to be dropped into the area where they were to be used. Hay bales would be dropped through side chutes, to be broken up and fed to the cattle. Straw bales would be dropped down the center chute, to be distributed as bedding in the livestock standing/resting areas.

Traditionally multiple bales were dropped down to the livestock floor and the twine removed by hand. After drying and being stored under tons of pressure in the haystack, most bales are tightly compacted and need to be torn apart and fluffed up for use.

One recent method of speeding up all this manual bale handling is the bale shredder, which is a large vertical drum with rotary cutting/ripping teeth at the base of the drum. The shredder is placed under the chute and several bales dropped in. A worker then pushes the shredder along the barn aisle as it rips up a bale and spews it out in a continuous fluffy stream of material.

Field of hay bales, lines in field made by baler.

Industrial Balers

Industrial balers are typically used to compact similar types of waste, such as office paper, corrugated fiberboard, plastic, foil and cans, for sale to recycling companies. These balers are made of steel with a hydraulic ram to compress the material loaded. Some balers are simple and labor-intensive, but are suitable for smaller volumes. Other balers are very complex and automated, and are used where large quantities of waste are handled.

A specialized baler designed to compact stretch wrap.

Used in recycling facilities, balers are a packaging step that allows for the aforementioned commodities to be broken down into dense cubes of one type of material at a time. There are different balers used depending on the material type. After a specific material is crushed down into a dense cube, it is tied to a bale by a thick wire and then pushed out of the machine. This process allows for easy transport of all materials involved.

- Two-ram baler: A two-ram baler is a baling machine that contains two cylinders and is able to bundle and package most commodities except for cardboard and clear film. This baler is known for its durability and is able to take in more bulky material.

- Single-ram baler: A single-ram baler is a baling machine that contains one cylinder. Because this baler is relatively smaller than the two-ram baler, it is best for small and medium commodities.

- Closed door baler: This baler bales clear plastic film.

- American baler: This baler bales corrugated materials.

HAY RAKE

A tractor with a 10-wheeled star-wheel rake forms a windrow.

A tractor with a rotary rake forms a windrow, another one with a loader wagon follows and collects the hay for silage.

A hay rake is an agricultural rake used to collect cut hay or straw into windrows for later collection (e.g. by a baler or a loader wagon). It is also designed to fluff up the hay and turn it over so that it may dry. It is also used in the evening to protect the hay from morning dew. The next day a tedder is used to spread it again, so that the hay dries more quickly.

Types

A hay rake may be mechanized, drawn by a tractor or draft animals, or it may be a hand tool. The earliest hay rakes were nothing more than tree branches, but wooden hand rakes with wooden teeth, similar in design to a garden rake but larger, were prevalent in the 19th and early 20th centuries, and still are used in some locations around the world.

The typical early horse-drawn hay rake was a *dump rake*, a wide two-wheeled implement with curved steel or iron teeth usually operated from a seat mounted over the rake with a lever-operated lifting mechanism. This rake gathered cut hay into windrows by repeated operation perpendicular to the windrow, requiring the operator to raise the rake, turn around and drop the teeth to rake back and forth in order to form the windrow. In some areas, a *sweep rake*, which could also be a horse-drawn or tractor-mounted implement, could then be used to pick up the windrowed hay and load it onto a wagon.

Later, a mechanically more complicated rake was developed, known as the *side delivery rake*. This usually had a gear-driven or chain-driven reel mounted roughly at a 45-degree angle to the windrow, so the hay was gathered and pushed to one side of the rake as it moved across the field. A side delivery rake could be pulled longitudinally along the windrow by horses or a tractor, eliminating the laborious and inefficient process of raising, lowering, and back-and-forth raking required by a dump rake. This allowed for the continuous spiraling windrows of a classic mid-20th-century farm hayfield. Later versions of the side delivery rake used a more severe transverse angle and a higher frame system, but the basic principles of operation were the same.

Still later, a variety of *wheel rakes* or *star wheel rakes* were developed, with 5, 6, 7 or more spring-tooth encircled wheels mounted on a frame and ground driven by free-wheeling contact as the implement was pulled forward. These rakes were variously promoted as being mechanically simpler and trouble-free, gentler on the hay than a side-delivery rake, and cheaper to operate.

Currently a newer design called the rotary rake is in common use in Europe, and less frequently seen in the United States and Canada.

A 19th-century hand-tool hay rake.

A late version of the side delivery rake.

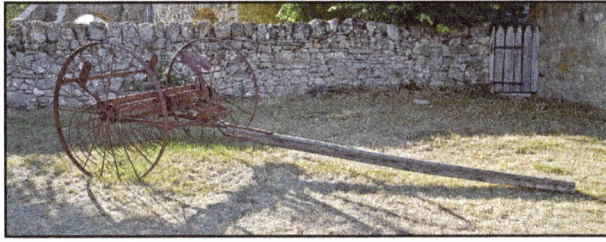

A dump rake.

TEDDER

A retired hay tedder.

A tedder (also called hay tedder) is a machine used in haymaking. It is used after cutting and before windrowing, and uses moving forks to aerate or "wuffle" the hay and thus speed up the process of hay-making. The use of a tedder allows the hay to dry ("cure") better, which results in improved aroma and color.

Operation

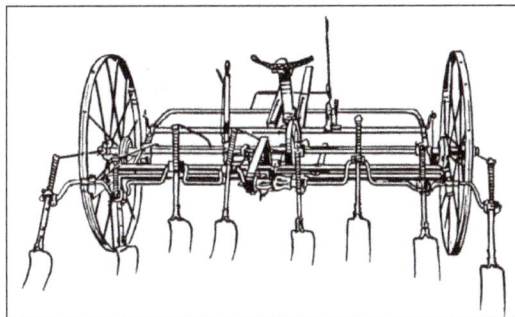

A hay tedder, similar to a standard American model of
the early 20th century, with tines shaped like pitchfork ends.

The original tedder is a farm tool on two wheels pulled by a horse; the rotation of the axle drives a gear which operates a "number of arms with wire tines or fingers at the lower ends." The tines pick up the hay and disperse it; usually, the height at which the tines pick up the hay can be adjusted.

In an early, simple hay tedder described in 1852 and manufactured in Edinburgh by the company of Mr. Slight, the two wheels, via a spur wheel and a pinion, drive a set of light wheels, the "rake wheels"; on these two rake wheels are mounted eight rakes, which pick up and disperse the hay. A later "English hay-tedder" uses two separate cylinders with rotating forks that can be reversed to lay the hay down lightly for improved exposure to air.

American machines, such as those made by Garfield, by Mudgett, and by Bullard typically used a system with a revolving crank in the middle of the arm and a lever at the upper end, or a system whereby rotating wheels moved the forks up and down. The first tedder widely available on the American market was the already mentioned Bullard's Hay Tedder, which had forks moving up and down on a compound crank, working in a motion described as "the energetic scratching of a hen." The American Hay Tedder, made by the Ames Plow Company of Boston and described in 1869 as a new machine, remarkable for its simplicity and perfection of working, was more like the British machine in its rotational operation.

Some tedders have the rotating tines enclosed inside a solid structure to increase the force applied to the hay. Other similar machines included *the wuffler* and *the acrobat*. The wuffler shuffles the hay in a manner similar to the tedder. The acrobat may be used also for turning, and for rowing hay up ready for baling.

Centrifugal Rakes

Tractor with rotary tedder.

On two opposing horizontal gyroscopes, which are pto-driven, are mounted obliquely downward standing tines. These refer to the green waste and throw it back. Due to the rear-mounted collecting baskets a windrowing is as possible with a Rake. Their distribution is low because of the limited job performance.

Use and Importance

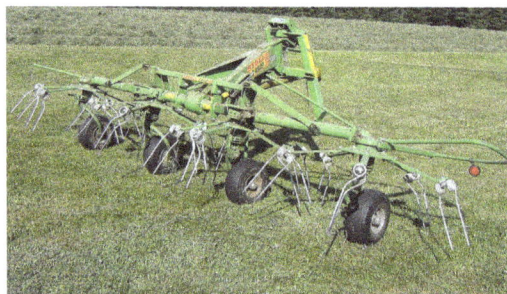
A modern tedder.

Its development was of great importance to agriculture, since it saved labor and thus money: using a tedder, a man and a horse could do as much work as fifteen laborers. It also resulted in greater economy, since cut grass could be turned into hay the same day even if it had become wet or been trampled by horses and before its nutritional value could be reduced by repeated soaking from rain. Especially in humid areas (such as the Eastern United States), the invention of the tedder added greatly to improved hay production from such crops as alfalfa and clover, and allowed for haying while the grass was still green which produced hay of much higher value.

CONDITIONER

A conditioner (or hay conditioner) is a farm implement that crimps and crushes newly cut hay to promote faster and more even drying. Drying the hay efficiently is most important for first cutting of the hay crop, which consists of coarse stalks that take a longer period of time to draw out moisture than finer textured hays, such as second and subsequent cuttings.

A conditioner is made up of two grooved rollers which the hay is forced through, causing the stalks to split, thus allowing the liquid trapped behind cell walls (sap and cell sap) to leak out and also giving more surface area for evaporation. The stand-alone conditioner is no longer used on most farms, since the conditioner has been incorporated into mower-conditioners, which combine the mower and conditioner into a single machine. The names Haybine and Discbine are brand names of mower-conditioners, although some farmers use these names somewhat generically.

Mower-conditioners

Mower with conditioner.

Mower-conditioners are a staple of large-scale hay making. Mower-conditioners are defined by the mechanisms that accomplish mowing and conditioning.

There are three types of mowers: sickle bar mowers, disc mowers, and drum mowers. Sickle bar mowers use a reciprocating blade to cut the grass and typically use a reel to fold the grass over the knife. Disc mowers have a number of hubs across the cutting width, each hub having a small (18") rotating disc with knives. Drum mowers use two or three large plates (called the drums, about 36" across) which ride over the ground as they are spinning. A sickle bar mower's main advantage over disc mowers and drum mowers is the reduced horsepower requirements. Its disadvantage is the extra maintenance required due to the high number of moving parts and wear items. Disc mowers were historically considered an "all the eggs in one basket" kind of mower because all the mower hubs were in one large gearbox. If one blade hit something and a gear tooth broke, the whole gearbox would suffer a catastrophic failure, and there would be nothing worth fixing. If anything broke, everything broke. Drum mowers prevented this by having typically two belt-driven drums compared to six or more gear-driven hubs. Modern disc mowers use isolated gearboxes, and if one fails it can be swapped out without rebuilding the entire machine.

Conditioners come in three main types: rubber-roller conditioners, steel-roller, and flail. The roller conditioners consist of two opposing rolls that have a raised, interlocking pattern. The rollers have either a rubber or steel pattern and a steel main shaft. The crop is crimped between the rollers, decreasing the drying time. The flail conditioner is an arrangement of steel V's on a main shaft that beat the crop against the top of the mower-conditioner. The flail conditioner reduces drying time by removing the waxy coating on the crop.

Even though conditioners can shorten the dry time of the hay, they can come with problems in the hayfield. The space between the two opposing rolls can decrease or increase by the users needs, but there is a max area of opening. This can cause a break down or problem in the machine. If the hay is wet or lay downed in the field it can bunch together when pushed into the conditioner rolls. This can jam the rolls, even causing the hay to wrap completely around them. If caught early, the user can shut down the machine and cut the hay free by hand. If it is not caught early this can stop the bars from rolling, thus stopping the belts from turning the machine, then stopping the tractor. If this happens, one is looking at major problems with their machinery. Not only can this damage the rolls, but also the belts on the mower-conditioner can snap and the tractor itself will be receiving considerable amounts of stress on its engine. Even though this can be a major money and time consuming problem, if the user is alert and constantly aware of their equipment's performance, this issue should never go further than a small stoppage.

Haybine is the brand name of the first mower-conditioner. It combined the sickle bar mower and the hay conditioner to promote faster drying hay all in one process. The current versions produced by New Holland are branded the *Discbine,* since they now feature faster disc mowers.

COTTON GIN

A cotton gin is a machine that quickly and easily separates cotton fibers from their seeds, enabling much greater productivity than manual cotton separation. The fibers are then processed into various cotton goods such as linens, while any undamaged cotton is used largely for textiles like clothing. The separated seeds may be used to grow more cotton or to produce cottonseed oil.

A modern mechanical cotton gin was created by American inventor Eli Whitney in 1793 and patented in 1794. Whitney's gin used a combination of a wire screen and small wire hooks to pull the cotton through, while brushes continuously removed the loose cotton lint to prevent jams. It revolutionized the cotton industry in the United States, but also led to the growth of slavery in the American South as the demand for cotton workers rapidly increased. The invention has thus been identified as an inadvertent contributing factor to the outbreak of the American Civil War. Modern automated cotton gins use multiple powered cleaning cylinders and saws, and offer far higher productivity than their hand-powered precursors.

A model of a 19th-century cotton gin on display at
the Eli Whitney Museum in Hamden, Connecticut.

Eli Whitney invented his cotton gin in 1793. He began to work on this project after moving to Georgia in search of work. Given that farmers were desperately searching for a way to make cotton farming profitable, a woman named Catharine Greene provided Whitney with funding to create the first cotton gin. Whitney created two cotton gins: a small one that could be hand-cranked and a large one that could be driven by a horse or water power.

Modern Cotton Gins

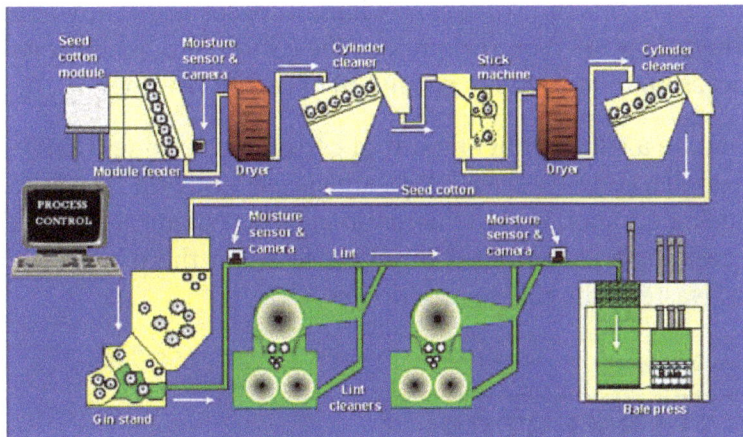

Diagram of a modern cotton gin plant,
displaying numerous stages of production.

In modern cotton production, cotton arrives at industrial cotton gins either in trailers, in compressed rectangular "modules" weighing up to 10 metric tons each or in polyethylene wrapped round modules similar to a bale of hay produced during the picking process by the most recent generation of cotton pickers. Cotton arriving at the gin is sucked in via a pipe, approximately 16 inches (41 cm) in diameter, that is swung over the cotton. This pipe is usually manually operated, but is increasingly automated in modern cotton plants. The need for trailers to haul the product to the gin has been drastically reduced since the introduction of modules. If the cotton is shipped in modules, the module feeder breaks the modules apart using spiked rollers and extracts the largest pieces of foreign material from the cotton. The module feeder's loose cotton is then sucked into the same starting point as the trailer cotton.

Modern cotton gins.

The cotton then enters a dryer, which removes excess moisture. The cylinder cleaner uses six or seven rotating, spiked cylinders to break up large clumps of cotton. Finer foreign material, such as soil and leaves, passes through rods or screens for removal. The stick machine uses centrifugal force to remove larger foreign matter, such as sticks and burrs, while the cotton is held by rapidly rotating saw cylinders.

The gin stand uses the teeth of rotating saws to pull the cotton through a series of "ginning ribs", which pull the fibers from the seeds which are too large to pass through the ribs. The cleaned seed is then removed from the gin via an auger conveyor system. The seed is reused for planting or is sent to an oil mill to be further processed into cottonseed oil and cottonseed meal. The lint cleaners again use saws and grid bars, this time to separate immature seeds and any remaining foreign matter from the fibers. The bale press then compresses the cotton into bales for storage and shipping. Modern gins can process up to 15 tonnes (33,000 lb) of cotton per hour.

Modern cotton gins create a substantial amount of cotton gin residue (CGR) consisting of sticks, leaves, dirt, immature bolls, and cottonseed. Research is currently under way to investigate the use of this waste in producing ethanol. Due to fluctuations in the chemical composition in processing, there is difficulty in creating a consistent ethanol process, but there is potential to further maximize the utilization of waste in the cotton production.

COMBINE HARVESTER

The combine harvester, or simply combine, is a machine that combines the tasks of harvesting, threshing, and cleaning grain crops.

The objective is the harvest of the crop; corn (maize), soybeans, flax (linseed), oats, wheat, or rye, among others. The waste straw left behind on the field is the remaining dried stems and leaves of the crop with limited nutrients which is either chopped and spread on the field or baled for feed and bedding for livestock.

A Lely open-cab combine.

The combine was patented in 1834, by Hiram Moore, the same year as Cyrus McCormick was granted a patent on the mechanical reaper.

Old Style Harverster found in the Henty, Australia region.

Early combines, some of them quite large, were drawn by horse or mule teams and used a bull wheel to provide power. Tractor-drawn, PTO-powered combines were used for a time. These combines used a shaker to separate the grain from the chaff and straw-walkers (grates with small teeth on an eccentric shaft) to eject the straw while retaining the grain. Tractor drawn combines evolved to have separate gas or diesel engines to power the grain separation. Today's combines are self-propelled and use diesel engines for power. A significant advance in the design of combines was the rotary design. Straw and grain were separated by use of a powerful fan. "Axial-Flow" rotary combines were introduced by International Harvester (IH) in 1977. About this time, on-board electronics were introduced to measure threshing efficiency. This new instrumentation allowed operators to get better grain yields by optimizing ground speed and other operating parameters.

Combine Heads

Combines are equipped with removable heads (called headers) that are designed for particular crops. The standard header, sometimes called a grain platform (or platform header), is equipped with a reciprocating knife cutter bar, and features a revolving reel with metal or plastic teeth to cause the cut crop to fall into the head. A cross auger then pulls the crop into the throat. The grain header is used for many crops, including grain, legumes, and many other crops.

Wheat headers are similar except that the reel is not equipped with teeth. Some wheat headers, called "draper" headers, use a fabric or rubber apron instead of a cross auger. Draper headers allow faster feeding than cross augers, leading to higher throughputs. In high yielding European crops, such headers have no advantage, as the limiting factor becomes grain separation. On many farms, platform headers are used to cut wheat, instead of separate wheat headers, so as to reduce overall costs.

A John Deere 9410 Combine set to harvest Oats.

A combine harvesting corn.

Dummy heads, or pick-up headers, feature spring-tined pickups, usually attached to a heavy rubber belt. They are used for crops that have already been cut and placed in windrows or swaths. This is particularly useful in northern climates, such as western Canada, where swathing kills weeds, resulting in a faster dry down.

While a grain platform can be used for corn, a specialized corn head is ordinarily used instead. The corn head is equipped with snap rolls that strip the stalk and leaf away from the ear, so that only the ear (and husk) enter the throat. This improves efficiency dramatically since so much less material must go through the cylinder. The corn head can be recognized by the presence of points between each row.

Occasionally, row-crop heads are seen that function like a grain platform, but have points between rows like a corn head. These are used to reduce the amount of weed seed picked up when harvesting small grains.

Self propelled Gleaner combines could be fitted with special tracks instead of tires to assist in harvesting rice. Some combines, particularly pull type, have tires with a diamond tread, which prevents sinking in mud.

Conventional Combine

The cut crop is carried up the feeder throat by a "chain and flight elevator," then fed into the threshing mechanism of the combine, consisting of a rotating threshing drum, to which grooved steel bars are bolted. These bars thresh or separate the grains and chaff from the straw through the action of the drum against the concave, a shaped "half drum," also fitted with steel bars and a meshed grill, through which grain, chaff and smaller debris may fall, whereas the straw, being too long, is carried through onto the straw walkers. The drum speed is variably adjustable, whilst the distance between the drum and concave is finely adjustable fore, aft and together, to achieve optimum separation and output. Manually engaged disawning plates are usually fitted to the concave. These provide extra friction to remove the awns from barley crops.

Sidehill Leveling

An interesting technology is in use in the Palouse region of the Pacific Northwest of the United States, in which the combine is retrofitted with a hydraulic sidehill leveling system. This allows the combine to harvest the incredibly steep but fertile soil in the region. Hillsides can be as steep as a 50 percent slope. Gleaner, IH and Case IH, John Deere, and others all have made combines with this sidehill leveling system, and local machine shops have fabricated them as an after-market add-on.

The first leveling technology was developed by Holt Co., a California firm, in 1891. Modern leveling came into being with the invention and patent of a level sensitive mercury switch system invented by Raymond Hanson in 1946. Raymond's son, Raymond, Jr., produced leveling systems exclusively for John Deere combines until 1995, as R. A. Hanson Company, Inc. In 1995, his son, Richard, purchased the company from his father and renamed it RAHCO International, Inc. In April, 2007, the company was renamed The Factory Company International, Inc. Production continues to this day.

Sidehill leveling has several advantages. Primary among them is an increased threshing efficiency on sidehills. Without leveling, grain and chaff slide to one side of separator and come through the machine in a large ball rather than being separated, dumping large amounts of grain on the ground. By keeping the machinery level, the straw-walker is able to operate more efficiently, making for more efficient threshing. IH produced the 453 combine, which leveled both side-to-side and front-to-back, enabling efficient threshing whether on a sidehill or climbing a hill head on.

Secondarily, leveling changes a combine's center of gravity relative to the hill and allows the combine to harvest along the contour of a hill without tipping, a very real danger on the steeper slopes of the region; it is not uncommon for combines to roll on extremely steep hills.

Currently, sidehill leveling is on the decline with the advent of huge modern machines which are more stable due to their width. These modern combines use the rotary grain separator which makes leveling less critical. Most combines on the Palouse have dual drive wheels on each side to stabilize them.

Maintaining Threshing Speed

Allis-Chalmers GLEANER L2.

CLAAS LEXION 570.

Another technology that is sometimes used on combines is a continuously variable transmission. This allows the ground speed of the machine to be varied while maintaining a constant engine and threshing speed. It is desirable to keep the threshing speed since the machine will typically have been adjusted to operate best at a certain speed.

Self-propelled combines started with standard manual transmissions that provided one speed based on input rpm. Deficiencies were noted, and in the early 1950s, combines were equipped with what John Deere called the "Variable Speed Drive." This was simply a variable width shive controlled by spring and hydraulic pressures. This shive was attached to the input shaft of the transmission. A standard 4 speed manual transmission was still used in this drive system. The operator would select a gear, typically third. An extra control was provided to the operator to allow him to speed up and slow down the machine within the limits provided by the variable speed drive system. By decreasing the width of the shive on the input shaft of the transmission, the belt would ride higher in the groove. This slowed the rotating speed on the input shaft of the transmission, thus slowing the ground speed for that gear. A clutch was still provided to allow the operator to stop the machine and change transmission gears.

Later, as hydraulic technology improved, hydrostatic transmissions were introduced by Versatile Mfg for use on swathers but later this technology was applied to combines as well. This drive retained the 4 speed manual transmission as before, but this time used a system of hydraulic pumps and motors to drive the input shaft of the transmission. This system is called a Hydrostatic drive system. The engine turns the hydraulic pump capable of high flow rates at up to 4000 psi. This pressure is then directed to the hydraulic motor that is connected to the input shaft of the transmission. The operator is provided with a lever in the cab that allows for the control of the hydraulic motors ability to use the energy provided by the pump. By adjusting the swash plate in the motor, the stroke of its pistons are changed. If the swash plate is set to neutral, the pistons do not move in their bores and no rotation is allowed, thus the machine does not move. By moving the lever, the swash plate moves its attached pistons forward, thus allowing them to move within the bore and causing the motor to turn. This provides an infinitely variable speed control from 0 ground speed to what ever the maximum speed is allowed by the gear selection of the transmission. The standard clutch was removed from this drive system, as it was no longer needed.

Most, if not all, modern combines are equipped with hydrostatic drives. These are larger versions of the same system used in consumer and commercial lawn mowers that most are familiar with today. In fact, it was the downsizing of the combine drive system that placed these drive systems into mowers and other machines.

Threshing Process

Despite great advances mechanically and in computer control, the basic operation of the combine harvester has remained unchanged almost since it was invented.

First of all the header, described above, cuts the crop and feeds it into the threshing cylinder. This consists of a series of horizontal rasp bars fixed across the path of the crop and in the shape of a quarter cylinder, guiding the crop upwards through a 90 degree turn. Moving rasp bars or rub bars pull the crop through concaved grates that separate the grain and chaff from the straw. The grain heads fall through the fixed concaves onto the sieves. The straw exits the top of the concave onto the straw walkers.

Since the IH 1440 and 1460 Axial-Flow Combines came out in 1977, combines have rotors in place of conventional cylinders. A rotor is a long, longitudinally mounted rotating cylinder with plates similar to rub bars.

There are usually two sieves, one above the other. Each is a flat metal plate with holes set according to the size of the grain mounted at an angle which shakes. The holes in the top sieve are set larger than the holes in the bottom sieve. While straw is carried to the rear, crop and weed seeds, as well as chaff, fall onto the second sieves, where chaff and crop fall though and are blown out by a fan. The crop is carried to the elevator which carries it into the hopper. Setting the concave clearance, fan speed, and sieve size is critical to ensure that the crop is threshed properly, the grain is clean of debris, and that all of the grain entering the machine reaches the grain tank. For example, when traveling uphill, the fan speed must be reduced to account for the shallower gradient of the sieves.

Heavy material, such as unthreshed heads, fall off the front of the sieves and are returned to the concave for re-threshing. The straw walkers are located above the sieves, and also have holes in them. Any grain remaining attached to the straw is shaken off and falls onto the top sieve.

When the straw reaches the end of the walkers it falls out the rear of the combine. It can then be baled for cattle bedding or spread by two rotating straw spreaders with rubber arms. Most modern combines are equipped with a straw spreader.

Rotary vs. Conventional Design

For a considerable time, combine harvesters used the conventional design, which used a rotating cylinder at the front-end which knocked the seeds out of the heads, and then used the rest of the machine to separate the straw from the chaff, and the chaff from the grain.

Case IH Combine set to harvest soybeans.

In the decades before the widespread adoption of the rotary combine in the late seventies, several inventors had pioneered designs which relied more on centrifugal force for grain separation and less on gravity alone. By the early eighties, most major manufacturers had settled on a "walkerless" design with much larger threshing cylinders to do most of the work. Advantages were faster grain harvesting and gentler treatment of fragile seeds, which were often cracked by the faster rotational speeds of conventional combine threshing cylinders.

The disadvantages of the rotary combine (which were increased power requirements and pulverization of the straw by-product) prompted a resurgence of conventional combines in the late nineties. Perhaps overlooked, but nonetheless true, when the large engines utilized to power the rotary machines were employed in conventional machines, the two types of machines delivered similar production capacities. Also, research was beginning to show that incorporating above-ground crop residue (straw) into the soil is less useful for rebuilding soil fertility than previously believed. This meant that working pulverized straw into the soil became more of a hindrance than a benefit. An increase in feedlot beef production also created a higher demand for straw as fodder. Conventional combines, which use straw walkers, preserve the quality of straw and allow it to be baled and removed from the field.

References

- Farm-machinery, technology: britannica.com, Retrieved 27 March 2019

- Wender, Charles H. (2004). Encyclopedia of American farm implements & antiques. Krause. P. 257. ISBN 978-0-87349-568-4

- Benefits-of-tractors-in-modern-farming-and-agriculture: farmmanagement.pro, Retrieved 28 April, 2019

- Bagg, Joel (2010-03-10). "Cutting, Conditioning & Raking For Faster Hay Drying". Ministry of Agriculture, Food and Rural Affairs: Crops; Field Crops; Fact Sheets. Ontario, CA: Queen's Printer for Ontario. Retrieved 20 May 2016. Joel Bagg, Forage Spacialist/OMAFRA

- Cultivator-farm-machine, technology: britannica.com, Retrieved 29 June, 2019

- Agblevor, Foster A.; Batz, Sandra; Trumbo, Jessica (February 3, 2018). "Composition and Ethanol Production Potential of Cotton Gin Residues". Biotechnology for Fuels and Chemicals. Humana Press, Totowa, NJ. Pp. 219–230. Doi:10.1007/978-1-4612-0057-4_17. ISBN 978-1-4612-6592-4

- Combine_harvester, entry: newworldencyclopedia.org, Retrieved 30 July, 2019

- Lee, Karen (2016-05-20). "Hay-in-a-Day: A Wide Swath More Important Than Conditioning". Information Center; Ag Practices and Trends. Brodhead, WI, USA: Kuhn North America, Inc. Retrieved 20 May 2016

PERMISSIONS

INDEX

www.ingramcontent.com/pod-product-compliance
Lightning Source LLC
Chambersburg PA
CBHW082025190326

41458CB00010B/3278